21st Century Homestead
Composting

Contents

Chapter 1

Bioeffector

A **Bioeffectors** is a viable microorganism or active natural compounds which directly or indirectly affects plant performance (Biofertilizer), and thus has the potential to reduce fertilizer and pesticide use in crop production.[*][1]

1.1 Types

Bioeffectors have a direct or indirect effect on plant performance by influencing the functional implementation or activation of biological mechanisms, particularly those interfering with soil-plant-microbe interactions.[*][2] In contrast to conventional fertilizers and pesticides, the effectiveness of bioeffectors is not based on a substantial direct input of mineral plant nutrients, either in inorganic or organic forms.

- Products in use are:
  - Microbial residues,
  - Composting and fermentation products,
  - Plant and algae extracts

- Bioeffector-preparations (*bio-agents*) as ready-formulated products are applied:
  - with the purpose of stimulating plant growth (bio-stimulants),
  - to improve plant nutrient acquisition (bio-fertilizers),
  - to protect plants from pathogens and pests (bio-control agents)
  - or generally to advance cropping efficiency; they can contain one or more bio-effectors along with other substances" [*][3]

- Well established bioeffectors with documented positive results in the field level are:
  - Rhizobia strains for soil or seed inoculation as a prerequisite for symbiotic N2-fixation when establishing new legume species or varieties.
  - positive effects of mycorrhiza inoculation for soils with a (temporarily) low potential for natural root mycorrhization.
  - sufficient mycorrhization enhances nutrient (P) and water uptake and increases resistance to pathogenic fungi.

- Further mechanisms for the positive impact of bioeffectors on plant growth have postulated, promising a high potential for resource preservation due to reduction of fertiliser and pesticide use:

- Active nutrient mobilisation by exudation of acids and carboxylates (e.g. P-mobilisation),

- exudation of micro-nutrient mobilising siderophores/chelates (e.g. Fe3+),

- reduction of trace elements from less soluble oxidised to highly soluble reduced forms (e.g. Fe3+ to Fe2+, Mn4+ to Mn2+),

- associative/non-symbiotic N2-fixation, protective antagonism to plant pathogens,

- enhancement of mycorrhizal infection and growth, and stimulating hormonal effects.

1.2   Research and Public Dissemination

Under the Acronym *Biofector* the European Union supports the Research of Bioeffectors under the leadership of the University of Hohenheim (Coordinator Guenter Neumann).[*][4] The results of the project will be evaluated by the members of the Association Biostimulants in Agriculture (ABISTA) and provided agriculture for use and EU institutions for the legislative and registration procedures.[*][5]

1.3   External links

- Webpage Biofector

- Webpage Association Biostimulants in Agriculture

- Webpage Biofector CULS Prague

- Webpage Madora Bioeffectors

- Webpage Biofector University of Hohenheim

1.4   References

[1] Minutes of the 6th International Symposium Plant Protection and Plant Health in Europe, May 2014 Braunschweig, Germany

[2] V. Römheld, G. Neumann (2006): *The Rhizosphere: Contributions of the soil-root interface to sustainable soil systems.* In: N. Uphoff, N., N. A. S. Ball et al. (Hg.), *Biological Approaches to Sustainable Soil Systems,* S. 92–107, CRC-Press, Oxford, UK.

[3] Bakonyi N., Donath S., Weinmann M., Neumann G., Müller T., Römheld V. (2008): Assessing commercial bio-fertilisers for improved phoshorus availability. Use of rapid screening tests. Jahrestagung der Deutschen Gesellschaft für Pflanzenernährung 2008

[4] European Research Program Biofector

[5] Webpage Biostimulants Association

Chapter 2

Compost

A community-level composting plant in a rural area in Germany

Compost (/ˈkɒmpɒst/ or /ˈkɒmpoʊst/) is organic matter that has been decomposed and recycled as a fertilizer and soil amendment. Compost is a key ingredient in organic farming. At the simplest level, the process of composting simply requires making a heap of wetted organic matter known as green waste (leaves, food waste) and waiting for the materials to break down into humus after a period of weeks or months. Modern, methodical composting is a multi-step, closely monitored process with measured inputs of water, air, and carbon- and nitrogen-rich materials. The decomposition process is aided by shredding the plant matter, adding water and ensuring proper aeration by regularly turning the mixture. Worms and fungi further break up the material. Bacteria requiring oxygen to function (aerobic bacteria) and fungi manage the chemical process by converting the inputs into heat, carbon dioxide and ammonium. The ammonium (NH_4) is the form of nitrogen used by plants. When available ammonium is not used by plants it is further converted by bacteria into nitrates (NO_3) through the process of nitrification.

Compost is rich in nutrients. It is used in gardens, landscaping, horticulture, and agriculture. The compost itself is beneficial for the land in many ways, including as a soil conditioner, a fertilizer, addition of vital humus or humic acids, and as a natural pesticide for soil. In ecosystems, compost is useful for erosion control, land and stream reclamation, wetland construction, and as landfill cover (see compost uses). Organic ingredients intended for composting can alternatively be used to generate biogas through anaerobic digestion.

2.1   Terminology

The term "composting" is used worldwide with differing meanings. Some composting textbooks narrowly define composting as being an aerobic form of decompostion, primarily by microbes. For many people, however, composting is used to refer to several different types of biological process. In North America, "anaerobic composting" is still a common term, but in much of the rest of the world and in technical publications the more accurate term anaerobic digestion is used as the microbes used and the processes involved are quite different.

2.2   Ingredients

Home compost barrel in the Escuela Barreales, Santa Cruz, Chile.

2.2.1   Carbon, nitrogen, oxygen, water

Materials in a compost pile.

Composting organisms require four equally important ingredients to work effectively:

- Carbon —for energy; the microbial oxidation of carbon produces the heat, if included at suggested levels.*[1]
  - High carbon materials tend to be brown and dry.
- Nitrogen —to grow and reproduce more organisms to oxidize the carbon.
  - High nitrogen materials tend to be green (or colorful, such as fruits and vegetables) and wet.*[2]
- Oxygen —for oxidizing the carbon, the decomposition process.
- Water —in the right amounts to maintain activity without causing anaerobic conditions.

Certain ratios of these materials will provide beneficial bacteria with the nutrients to work at a rate that will heat up the pile. In that process much water will be released as vapor ("steam"), and the oxygen will be quickly depleted, explaining the need to actively manage the pile. The hotter the pile gets, the more often added air and water is necessary; the air/water balance is critical to maintaining high temperatures (135°−160° Fahrenheit / 50° - 70° Celsius) until the materials are broken down. At the same time, too much air or water also slows the process, as does too much carbon (or too little nitrogen).

The most efficient composting occurs with an optimal carbon:nitrogen ratio of about 10:1 to 20:1.*[3] Nearly all plant and animal materials have both carbon and nitrogen, but amounts vary widely, with characteristics noted above (dry/wet,

Food scraps compost heap.

brown/green).[4] Fresh grass clippings have an average ratio of about 15:1 and dry autumn leaves about 50:1 depending on species. Mixing equal parts by volume approximates the ideal C:N range. Few individual situations will provide the ideal mix of materials at any point. Observation of amounts, and consideration of different materials[5] as a pile is built over time, can quickly achieve a workable technique for the individual situation.

2.2.2 Animal manure and bedding

On many farms, the basic composting ingredients are animal manure generated on the farm and bedding. Straw and sawdust are common bedding materials. Non-traditional bedding materials are also used, including newspaper and chopped cardboard. The amount of manure composted on a livestock farm is often determined by cleaning schedules, land availability, and weather conditions. Each type of manure has its own physical, chemical, and biological characteristics. Cattle and horse manures, when mixed with bedding, possess good qualities for composting. Swine manure, which is very wet and usually not mixed with bedding material, must be mixed with straw or similar raw materials. Poultry manure also must be blended with carbonaceous materials - those low in nitrogen preferred, such as sawdust or straw.[6]

2.2.3 Microorganisms

With the proper mixture of water, oxygen, carbon, and nitrogen, micro-organisms are allowed to break down organic matter to produce compost. The composting process is dependent on micro-organisms to break down organic matter into compost. There are many types of microorganisms found in active compost of which the most common are:[7]

- Bacteria- The most numerous of all the microorganisms found in compost. Depending on the phase of composting, mesophilic or thermophilic bacteria may predominate.

- Actinobacteria- Necessary for breaking down paper products such as newspaper, bark, etc.

- Fungi- Molds and yeast help break down materials that bacteria cannot, especially lignin in woody material.

- Protozoa- Help consume bacteria, fungi and micro organic particulates.

- Rotifers- Rotifers help control populations of bacteria and small protozoans.

In addition, earthworms not only ingest partly composted material, but also continually re-create aeration and drainage tunnels as they move through the compost.

A lack of a healthy micro-organism community is the main reason why composting processes are slow in landfills with environmental factors such as lack of oxygen, nutrients or water being the cause of the depleted biological community.[*][7]

Phases of composting

Under ideal conditions, composting proceeds through three major phases:[*][7]

- An initial, mesophilic phase, in which the decomposition is carried out under moderate temperatures by mesophilic microorganisms.

- As the temperature rises, a second, thermophilic phase starts, in which in decomposition is carried out by various thermophilic bacteria under high temperatures.

- As the supply of high-energy compounds dwindles, the temperature starts to decrease, and the mesophiles once again predominate in the maturation phase.

2.2.4 Human waste

Human waste (excreta) can also be added as an input to the composting process, like it is done in composting toilets, as human waste is a nitrogen-rich organic material.

People excrete far more water-soluble plant nutrients (nitrogen, phosphorus, potassium) in urine than in feces.[*][8] Human urine can be used directly as fertilizer or it can be put onto compost. Adding a healthy person's urine to compost usually will increase temperatures and therefore increase its ability to destroy pathogens and unwanted seeds. Urine from a person with no obvious symptoms of infection is much more sanitary than fresh feces. Unlike feces, urine does not attract disease-spreading flies (such as house flies or blow flies), and it does not contain the most hardy of pathogens, such as parasitic worm eggs. Urine usually does not stink for long, particularly when it is fresh, diluted, or put on sorbents.

Urine is primarily composed of water and urea. Although metabolites of urea are nitrogen fertilizers, it is easy to over-fertilize with urine, or to utilize urine containing pharmaceutical (or other) content, creating too much ammonia for plants to absorb, acidic conditions, or other phytotoxicity.[*][9]

Humanure

"Humanure" is a combination of the words *human* and *manure*, designating human excrement (feces and urine) that is recycled via composting for agricultural or other purposes. The term was first used in a 1994 book by Joseph Jenkins that advocates the use of this organic soil amendment.[*][10] The term humanure is used by compost enthusiasts in the US but not generally elsewhere. Because the term "humanure" has no authoritative definition it is subject to various uses; news reporters occasionally fail to correctly distinguish between humanure and sewage sludge or "biosolids".[*][11]

2.3 Uses

Main article: Uses of compost

Compost is generally recommended as an additive to soil, or other matrices such as coir and peat, as a tilth improver, supplying humus and nutrients. It provides a rich *growing medium*, or a porous, absorbent material that holds moisture and soluble minerals, providing the support and nutrients in which plants can flourish, although it is rarely used alone, being primarily mixed with soil, sand, grit, bark chips, vermiculite, perlite, or clay granules to produce loam. Compost can be tilled directly into the soil or growing medium to boost the level of organic matter and the overall fertility of the soil. Compost that is ready to be used as an additive is dark brown or even black with an earthy smell.*[12]

Generally, direct seeding into a compost is not recommended due to the speed with which it may dry and the possible presence of phytotoxins that may inhibit germination,*[13]*[14]*[15] and the possible tie up of nitrogen by incompletely decomposed lignin.*[5] It is very common to see blends of 20–30% compost used for transplanting seedlings at cotyledon stage or later.

Composting can destroy pathogens or unwanted seeds. Unwanted living plants (or weeds) can be discouraged by covering with mulch/compost. The "microbial pesticides" in compost may include thermophiles and mesophiles, however certain composting detritivores such as black soldier fly larvae and redworms, also reduce many pathogens. Thermophilic (high-temperature) composting is well known to destroy many seeds and nearly all types of pathogens (exceptions may include prions). The sanitizing qualities of (thermophilic) composting are desirable where there is a high likelihood of pathogens, such as with manure.

2.4   Composting technologies

A homemade compost tumbler

A modern compost bin constructed from plastics

2.4.1 Overview

In addition to the traditional compost pile, various approaches have been developed to handle different composting processes, ingredients, locations, and applications for the composted product.

There is a large number of different composting systems on the market, for example:

- At the household level: Composting toilet, container composting, vermicomposting

- At the industrial composting (large scale): Aerated Static Pile Composting, vermicomposting, windrow composting etc.

2.4.2 Examples

Vermicomposting

Main article: Vermicomposting

Vermicompost is the product or process of composting through the utilization of various species of worms, usually red wigglers, white worms, and earthworms, to create a heterogeneous mixture of decomposing vegetable or food waste (excluding meat, dairy, fats, or oils), bedding materials, and vermicast. Vermicast, also known as worm castings, worm humus or worm manure, is the end-product of the breakdown of organic matter by species of earthworm.[16] Vermicomposting is widely used in North America for on-site institutional processing of food waste, such as in hospitals and

Rotary screen harvested worm castings

shopping malls. This type of composting is sometimes suggested as a feasible indoor home composting method. Vermi-composting has gained popularity in both these industrial and domestic settings because, as compared with conventional composting, it provides a way to compost organic materials more quickly (as defined by a higher rate of carbon-to-nitrogen ratio increase) and to attain products that have lower salinity levels that are therefore more beneficial to plant mediums.[*][17]

Food waste - after three years

The earthworm species (or **composting worms**) most often used are red wigglers (*Eisenia fetida* or *Eisenia andrei*), though European nightcrawlers (*Eisenia hortensis* or *Dendrobaena veneta*) could also be used. Red wigglers are recommended by most vermiculture experts, as they have some of the best appetites and breed very quickly. Users refer to European nightcrawlers by a variety of other names, including *dendrobaenas*, *dendras*, Dutch Nightcrawlers, and Belgian nightcrawlers.

Containing water-soluble nutrients, vermicompost is a nutrient-rich organic fertilizer and soil conditioner in a form that is relatively easy for plants to absorb.[18] Worm castings are sometimes used as an organic fertilizer. Because the earthworms grind and uniformly mix minerals in simple forms, plants need only minimal effort to obtain them. The worms' digestive systems also add beneficial microbes to help create a "living" soil environment for plants.

Vermicompost tea in conjunction with 10% castings has been shown to cause up to a 1.7 times growth in plant mass over plants grown without.[19]

Researchers from the Pondicherry University discovered that worm composts can also be used to clean up heavy metals. The researchers found substantial reductions in heavy metals when the worms were released into the garbage and they are effective at removing lead, zinc, cadmium, copper and manganese.[20]

Hügelkultur (raised garden beds or mounds)

Main article: Hügelkultur

The practice of making raised garden beds or mounds filled with rotting wood is also called "Hügelkultur" in Ger-

An almost completed Hügelkultur bed (does not have dirt on it yet).

man.[21][22] It is in effect creating a Nurse log, however, covered with dirt.

Benefits of hügelkultur garden beds include water retention and warming of soil.[21][23] Buried wood becomes like a sponge as it decomposes, able to capture water and store it for later use by crops planted on top of the hügelkultur bed.[21][24]

The buried decomposing wood will also give off heat, as all compost does, for several years. These effects have been used by Sepp Holzer for one to allow fruit trees to survive at otherwise inhospitable temperatures and altitudes.[22]

Black soldier fly larvae composting

Main article: Hermetia illucens § Uses in composting or as food for animals

Black Soldier Fly (*Hermetia illucens*) larvae have been shown to be able to rapidly consume large amounts of organic waste when kept at 31.8°C, the optimum temperature for reproduction. [25] Enthusiasts have experimented with a large number of different waste products[26] and some even sell starter kits to the public.[27]

Cockroach composting

Cockroach composting is another insect-mediated composting method. In this case the adults of any number of cockroach species (such as the Turkestan cockroach or *Blaptica dubia*) are used to quickly convert manure or kitchen waste to nutrient dense compost. Depending on species used and environmental conditions, excess composting insects can be used as an excellent animal feed for farm animals and pets.[28]

Bokashi

Bokashi is a method that uses a mix of microorganisms to cover food waste to decrease smell. It derives from the practice of Japanese farmers centuries ago of covering food waste with rich, local soil that contained the microorganisms that would ferment the waste. After a few weeks, they would bury the waste. [29]

Most practitioners obtain the microorganisms from the product Effective Microorganisms (EM1),[29] first sold in the 1980s. EM1 is mixed with a carbon base (e.g. sawdust or bran) that it sticks to and a sugar for food (e.g. molasses). The mixture is layered with waste in a sealed container and after a few weeks, removed and buried.[29]

Newspaper fermented in a lactobacillus culture can be substituted for bokashi bran for a successful bokashi bucket. [30]

Compost tea

Compost teas are defined as water extracts brewed from composted materials and can be derived from aerobic or anaerobic processes.[31] Compost teas are generally produced from adding one volume of compost to 4-10 volumes of water, but there has also been debate about the benefits of aerating the mixture.[31] Field studies have shown the benefits of adding compost teas to crops due to the adding of organic matter, increased nutrient availability and increased microbial activity.[31] They have also been shown to have an effect on plant pathogens.[32]

Composting toilets

Main article: Composting toilet

A composting toilet does not require water or electricity, and when properly managed does not smell. A composting toilet collects human excreta which is then added to a compost heap together with sawdust and straw or other carbon rich materials, where pathogens are destroyed to some extent. The amount of pathogen destruction depends on the temperature (mesophilic or thermophilic conditions) and composting time.[33] A composting toilet tries to process the excreta in situ

Inside a recently started bokashi bin. The aerated base is just visible through the food scraps and bokashi bran.

although this is often coupled with a secondary external composting step. The resulting compost product has been given various names, such as humanure and EcoHumus.*[33]

A composting toilet can aid in the conservation of fresh water by avoiding the usage of potable water required by the typical flush toilet. It further prevents the pollution of ground water by controlling the fecal matter decomposition before entering the system. When properly managed, there should be no ground contamination from leachate.

2.5   Compost and land-filling

As concern about landfill space increases, worldwide interest in recycling by means of composting is growing, since composting is a process for converting decomposable organic materials into useful stable products.*[34] Composting is one of the only ways to revitalize soil vitality due to phosphorus depletion in soil.*[35] Industrial scale composting in the form of in-vessel composting, aerated static pile composting, and anaerobic digestion takes place in most Western countries now, and in many areas is mandated by law. There are process and product guidelines in Europe that date to the early 1980s (Germany, the Netherlands, Switzerland) and only more recently in the UK and the US. In both these countries, private trade associations within the industry have established loose standards, some say as a stop-gap measure to discourage independent government agencies from establishing tougher consumer-friendly standards.*[36] The USA is the only Western country that does not distinguish sludge-source compost from green-composts, and by default in the USA 50% of states expect composts to comply in some manner with the federal EPA 503 rule promulgated in 1984 for sludge products.*[37] Compost is regulated in Canada and Australia as well.

2.5.1 Industrial systems

A large compost pile that is steaming with the heat generated by thermophilic microorganisms.

Industrial composting systems are increasingly being installed as a waste management alternative to landfills, along with other advanced waste processing systems. Mechanical sorting of mixed waste streams combined with anaerobic digestion or in-vessel composting is called mechanical biological treatment, and are increasingly being used in developed countries due to regulations controlling the amount of organic matter allowed in landfills. Treating biodegradable waste before it enters a landfill reduces global warming from fugitive methane; untreated waste breaks down anaerobically in a landfill, producing landfill gas that contains methane, a potent greenhouse gas.

Vermicomposting, also known as vermiculture, is used for medium-scale on-site institutional composting, such as for food waste from universities and shopping malls: selected either as a more environmental choice, or to reduce the cost of commercial waste removal.

Large-scale composting systems are used by many urban areas around the world. Co-composting is a technique that combines solid waste with de-watered biosolids, although difficulties controlling inert and plastics contamination from municipal solid waste makes this approach less attractive. The World's largest MSW co-composter is the Edmonton Composting Facility in Edmonton, Alberta, Canada, which turns 220,000 tonnes of residential solid waste and 22,500 dry tonnes of biosolids per year into 80,000 tonnes of compost. The facility is 38,690 meters^2 (416,500 ft^2), equivalent to 4½ Canadian football fields, and the operating structure is the largest stainless steel building in North America, the size of 14 NHL rinks. In 2006, the State of Qatar awarded Keppel Seghers Singapore, a subsidiary of Keppel Corporation to begin construction on a 275,000 tonne/year Anaerobic Digestion and Composting Plant licensed by Kompogas Switzerland. This plant, with 15 independent anaerobic digestors will be the world's largest composting facility once fully operational in early 2011 and forms part of the Qatar Domestic Solid Waste Management Center, the largest integrated waste management complex in the Middle East.

Another large MSW composter is the Lahore Composting Facility in Lahore, Pakistan, which has a capacity to convert 1,000 tonnes of municipal solid waste per day into compost. It also has a capacity to convert substantial portion of the intake into Refuse-derived fuel (RDF) materials for further combustion use in several energy consuming industries across

Pakistan e.g., in cement manufacturing companies where it is used to heat up the Cement Kiln systems. This project has also been approved by the Executive Board of the United Nations Framework Convention on Climate Change (UNFCCC) for reduction of emission of methane gas into the climate and has been registered with a capacity of reducing 108,686 metric tonnes CO2 equivalent per annum.*[38]

2.6   Related technologies

Anaerobic digestion is another possible process for converting organic waste into a useful produce (biogas). In central Europe, anaerobic digestion is now more common than composting as a process for treating organic waste. The two processes can also be used in combination: sewage sludge is often anaerobically digested first, followed by a composting process before selling or giving away the compost to farmers.

2.7   History

Composting as a recognized practice dates to at least the early Roman Empire since Pliny the Elder (AD 23-79). Traditionally, composting involved piling organic materials until the next planting season, at which time the materials would have decayed enough to be ready for use in the soil. The advantage of this method is that little working time or effort is required from the composter and it fits in naturally with agricultural practices in temperate climates. Disadvantages (from the modern perspective) are that space is used for a whole year, some nutrients might be leached due to exposure to rainfall, and disease-producing organisms and insects may not be adequately controlled.

Composting was somewhat modernized beginning in the 1920s in Europe as a tool for organic farming. The first industrial station for the transformation of urban organic materials into compost was set up in Wels, Austria in the year 1921.*[39] Early frequent citations for propounding composting within farming are for the German-speaking world Rudolf Steiner, founder of a farming method called biodynamics, and Annie Francé-Harrar, who was appointed on behalf of the government in Mexico and supported the country 1950–1958 to set up a large humus organization in the fight against erosion and soil degradation. In the English-speaking world it was Sir Albert Howard who worked extensively in India on sustainable practices and Lady Eve Balfour who was a huge proponent of composting. Composting was imported to America by various followers of these early European movements by the likes of J.I. Rodale (founder of Rodale Organic Gardening), E.E. Pfeiffer (who developed scientific practices in biodynamic farming), Paul Keene (founder of Walnut Acres in Pennsylvania), and Scott and Helen Nearing (who inspired the back-to-the-land movement of the 1960s). Coincidentally, some of the above met briefly in India - all were quite influential in the U.S. from the 1960s into the 1980s.

There are many modern proponents of rapid composting that attempt to correct some of the perceived problems associated with traditional, slow composting. Many advocate that compost can be made in 2 to 3 weeks.*[40] Many such short processes involve a few changes to traditional methods, including smaller, more homogenized pieces in the compost, controlling carbon-to-nitrogen ratio (C:N) at 30 to 1 or less, and monitoring the moisture level more carefully. However, none of these parameters differ significantly from the early writings of Howard and Balfour, suggesting that in fact modern composting has not made significant advances over the traditional methods that take a few months to work. For this reason and others, many modern scientists who deal with carbon transformations are sceptical that there is a "super-charged" way to get nature to make compost rapidly.

In fact, both sides are right to some extent. The bacterial activity in rapid high heat methods breaks down the material to the extent that pathogens and seeds are destroyed, and the original feedstock is unrecognizable. At this stage, the compost can be used to prepare fields or other planting areas. However, most professionals recommend that the compost be given time to cure before using in a nursery for starting seeds or growing young plants. The curing time allows fungi to continue the decomposition process and eliminating phytotoxic substances.

Many countries such as Wales*[41]*[42] and some individual cities such as Seattle and San Francisco require food and yard waste to be sorted for composting.*[43]*[44]

Kew Gardens in London has one of the biggest non-commercial compost heaps in Europe.

2.8 See also

- List of composting systems

- San Francisco Mandatory Recycling and Composting Ordinance

- Terra preta

- Urban agriculture

- Waste sorting

2.9 References

[1] "Composting for the Homeowner - University of Illinois Extension". Web.extension.illinois.edu. Retrieved 2013-07-18.

[2] "Composting for the Homeowner - University of Illinois Extension". *uiuc.edu*.

[3] Radovich, T; Hue, N; Pant, A (2011). "Chapter 1: Compost Quality". In Radovich, T; Arancon, N. *Tea Time in the Tropics - a handbook for compost tea production and use*. College of Tropical Agriculture and Human Resources, University of Hawaii. pp. 8–16. External link in |title= (help)

[4] Klickitat County WA, USA Compost Mix Calculator

[5] "The Effect of Lignin on Biodegradability - Cornell Composting". *cornell.edu*.

[6] Dougherty, Mark. (1999). Field Guide to On-Farm Composting. Ithaca, New York: Natural Resource, Agriculture, and Engineering Service.

[7] "Composting - Compost Microorganisms". *Cornell University*. Retrieved 6 October 2010.

[8] Stockholm Environment Institute - EcoSanRes - Guidelines on the Use of Urine and Feces in Crop Production

[9] "TUBdok: Pharmaceutical Residues in Urine and Potential Risks related to Usage as Fertiliser in Agriculture" (PDF). *tu-harburg.de*.

[10] Jenkins, J.C. (2005). *The Humanure Handbook: A Guide to Composting Human Manure*. Grove City, PA: Joseph Jenkins, Inc.; 3rd edition. p. 255. ISBN 978-0-9644258-3-5. Retrieved April 2011.

[11] Courtney Symons (13 October 2011). "'Humanure' dumping sickens homeowner". *YourOttawaRegion*. Metroland Media Group Ltd. Retrieved 16 October 2011.

[12] Healthy Soils, Healthy Landscapes

[13] Morel, P. and Guillemain, G. 2004. Assessment of the possible phytotoxicity of a substrate using an easy and representative biotest. Acta Horticulture 644:417–423

[14] Itävaara et al. Compost maturity - problems associated with testing. in Proceedings of Composting. Innsbruck Austria 18-21.10.2000

[15] Aslam DN, et al. "Development of models for predicting carbon mineralization and associated phytotoxicity in compost-amended soil.". *nih.gov*.

[16] "Paper on Invasive European Worms". Retrieved 22 February 2009.

[17] Lazcano, Cristina; Gómez-Brandón, María; Domínguez, Jorge (2008). "Comparison of the effectiveness of composting and vermicomposting for the biological stabilization of cattle manure" (PDF). *Chemosphere* **72**: 1013–1019. doi:10.1016/j.chemosphere.2008.04.016.

[18] Coyne, Kelly and Erik Knutzen. *The Urban Homestead: Your Guide to Self-Sufficient Living in the Heart of the City*. Port Townsend: Process Self Reliance Series, 2008.

[19] "Worm casting organic fertilizer benefits and uses". *Worms Etc*.

18

CHAPTER 2. COMPOST

[20] *Cleaning up heavy metals using worms*, International: mining.com, 2012, retrieved 3 October 2012

[21] "hugelkultur: the ultimate raised garden beds". Richsoil.com. 2007-07-27. Retrieved 2013-07-18.

[22] "The Art and Science of Making a Hugelkultur Bed - Transforming Woody Debris into a Garden Resource Permaculture Research Institute - Permaculture Forums, Courses, Information & News". Retrieved 2013-07-18.

[23] "Hugelkultur: Composting Whole Trees With Ease Permaculture Research Institute - Permaculture Forums, Courses, Information & News". Retrieved 2013-07-18.

[24] Hemenway, Toby (2009). Gaia's Garden: A Guide to Home-Scale Permaculture. Chelsea Green Publishing. pp. 84-85. ISBN 978-1-60358-029-8.

[25] Diener, Stefan; Studt Solano, Nandayure M.; Roa Gutiérrez, Floria; Zurbrügg, Christian; Tockner, Klement (2011). "Biological Treatment of Municipal Organic Waste using Black Soldier Fly Larvae". *Waste and Biomass Valorization* **2** (4): 357–363. doi:10.1007/s12649-011-9079-1. ISSN 1877-2641.

[26] "E". *Bio-Conversion of Putrescent Waste*. ESR International. Retrieved 17 April 2015.

[27] "BSF Farming - marketplace". Retrieved 17 April 2015.

[28] "Cockroach Composting". *The Unconventional Farmer*.

[29] Lindsay, Jay (12 June 2012). "Japanese composting may be new food waste solution". *AP*. Retrieved 13 November 2012.

[30] "Make your own FREE bokashi starter", 12 September 2008. Retrieved 7 November 2013.

[31] Gómez-Brandón, M; Vela, M; Martinez Toledo, MV; Insam, H; Domínguez, J (2015). "12: Effects of Compost and Vermicompost Teas as Organic Fertilizers". In Sinha, S; Plant, KK; Bajpai, S. *Advances in Fertilizer Technology: Synthesis (Vol1)*. Stadium Press LLC. pp. 300–318. ISBN 1-62699-044-1.

[32] Santos, M; Dianez, F; Carretero,F (2011). "12: Suppressive Effects of Compost Tea on Phytopathogens". In Dubey,NK. *Natural products in plant pest management*. Oxfordshire, UK Cambridge, MA: CABI. pp. 242–262. ISBN 9781845936716.

[33] Stenström, T.A., Seidu, R., Ekane, N., Zurbrügg, C. (2011). Microbial exposure and health assessments in sanitation technologies and systems - EcoSanRes Series, 2011-1. Stockholm Environment Institute (SEI), Stockholm, Sweden, page 88

[34] A Brief History of Solid Waste Management

[35] "Preventing Contaminants in Home Compost Piles". Retrieved 16 June 2012.

[36] "US Composting Council". Compostingcouncil.org. Retrieved 2013-07-18.

[37] "Electronic Code of Federal Regulations. Title 40, part 503. Standards for the use or disposal of sewage sludge". *U.S. Government Printing Office*. 1998. Retrieved 30 March 2009.

[38] Details on project design and its validation and monitoring reports are available at: Project 2778 : Composting of Organic Content of Municipal Solid Waste in Lahore

[39] Welser Anzeiger vom 05. Januar 1921, 67. Jahrgang, Nr. 2, S. 4

[40] The Rapid Compost Method by Robert Raabe, Professor of Plant Pathology, Berkeley

[41] Gwynedd Council food recycling

[42] "Anglesey households achieve 100% food waste recycling". *edie.net*.

[43] "San Francisco Signs Mandatory Recycling & Composting Laws". Retrieved 19 September 2010.

[44] Tyler, Aubin (21 March 2010). "The case for mandatory composting". *The Boston Globe*. Retrieved 19 September 2010.

Chapter 3

Hügelkultur

Completed Hügelkultur bed prior to being covered with soil.

Hügelkultur is a composting process employing raised planting beds constructed on top of decaying wood debris and other compostable biomass plant materials. The process helps to improve soil fertility, water retention, and soil warming, thus benefiting plants grown on or near such mounds.*[1]*[2]

3.1   Description

Hügelkultur replicates the natural process of decomposition that occurs on forest floors. Trees that fall in a forest often become nurse logs*[3] decaying and providing ecological facilitation to seedlings. As the wood decays, its porosity

Hügelkultur bed with wildflower overplanting

increases allowing it to store water "like a sponge" . The water is slowly released back into the environment, benefiting nearby plants.[1]

Mounded hügelkultur beds are ideal for areas where the underlying soil is of poor quality or compacted. They tend to be easier to maintain due to their relative height above the ground.[3] The beds are usually about 3 feet (0.91 m) by 6 feet (1.8 m) in area and about 3 feet (0.91 m) high.[1]

3.2 History

Hügelkultur is German word meaning mound culture or hill culture.[4] It was practiced in German and Eastern European culture for hundreds of years,[1][5] before being further developed by Sepp Holzer, an Austrian permaculture expert.[3] In addition, recent permaculture voices such as Paul Wheaton and Geoff Lawton advocate strongly for Hügelkultur beds as a perfect permaculture design.[6]

3.3 Construction

In its basic form, mounds are constructed by piling logs, branches, plant waste, compost and additional soil directly on the ground or in a shallow swale.[7][8] Mounds may also be made from alternating layers of wood, sod,[9] compost, straw, and soil. Although their construction is straightforward, planning is necessary to prevent steep slopes that would result in erosion.[3][5] Some designs recommend that mounds have a grade of between 65 and 80 degrees.[10]

Hügelkultur bed construction, shown without the top layer of soil

In his book *Desert or Paradise: Restoring Endangered Landscapes Using Water Management, Including Lake and Pond Construction*, Holzer describes a method of constructing Hügelkultur including a design that incorporates rubbish such as cardboard, clothes and kitchen waste. He recommends building mounds that are 1 meter (3.3 ft) wide and any length. Mounds are built in a 0.7 meters (2.3 ft) trench in sandy soil, and without a trench if the ground is wet.*[10]

3.4   See Also

- Sepp Holzer

- Paul Wheaton

- Permaculture

3.5   References

[1] Miles, Melissa (August 3, 2010). "The Art and Science of Making a Hugelkultur Bed – Transforming Woody Debris into a Garden Resource". The Permaculture Research Institute. Retrieved May 2, 2014.

[2] "The Many Benefits of Hugelkultur". Permaculture Magazine. October 17, 2013. Retrieved May 2, 2014.

[3] Palmer, Kim (August 14, 2013). "A Garden Made of WOOD; Hugelkultur (Hooogellocullocher or Hewogellocullocher) A Nature-Inspired Method of Gardening in Beds Built on Logs, Touted as a Drought-Resistant Way to Produce Food". Minneapolis, MN. Star Tribune. Retrieved May 2, 2014.

[4] Lauterbach, Margaret (February 2, 2012). "Margaret Lauterbach: Clippings fuel fertile 'hugel' mounds". Boise, ID. Idaho Statesman. Retrieved May 3, 2014.

[5] Martin, Claire (April 11, 2014). "Hugelkultur , translated: A path to richer soil". Denver, CO. Denver Post. Retrieved May 2, 2014.

[6] Wheaton, Paul, hugelkultur: the ultimate raised garden beds, retrieved 6/9/2014

[7] Hemenway, Toby (2009). *Gaia's Garden: A Guide to Home-Scale Permaculture*. Chelsea Green Publishing. pp. 84–85. ISBN 9781603582230.

[8] Feineigle, Mark (January 4, 2012). "Hugelkultur: Composting Whole Trees With Ease". The Permaculture Research Institute. Retrieved May 2, 2014.

[9] Shein, Christopher (2013). *The Vegetable Gardener's Guide to Permaculture: Creating an Edible Ecosystem*. Timber Press. p. 28. ISBN 978-1604692709.

[10] Holzer, Sepp (2012). *Hügelkultur*. Chelsea Green Publishing. pp. 131–134, 139. ISBN 978-1603584647.

3.6   External links

- Hugelkultur: The Ultimate Raised Garden Beds by Paul Wheaton

- Hugelkultur Video

- hugelkultur forum on Hugelkultur at permies.com

Chapter 4

List of composting systems

A modern compost bin constructed from plastics

The following page contains a **list of different composting systems**:

4.1 Home composting (small-scale)

- Composting toilet

- Container composting

- Ecuador composting method

- German mound

- Sheet composting

- Trench composting

- Vermicomposting

4.2   Industrial composting (large scale)

Aeration system for a closed chamber composting facility

- Aerated Static Pile Composting

- High fibre composting

- In-vessel composting

- Mechanical biological treatment

- Tunnel composting

- Vermicomposting
- Windrow composting

4.3   See also

- Compost

Chapter 5

Aerated static pile composting

Aerated Static Pile (ASP) composting, refers to any of a number of systems used to biodegrade organic material without physical manipulation during primary composting. The blended admixture is usually placed on perforated piping, providing air circulation for controlled aeration . It may be in windrows, open or covered, or in closed containers. With regard to complexity and cost, aerated systems are most commonly used by larger, professionally managed composting facilities, although the technique may range from very small, simple systems to very large, capital intensive, industrial installations.*[1]

Aerated static piles offer process control for rapid biodegradation, and work well for facilities processing wet materials and large volumes of feedstocks. ASP facilities can be under roof or outdoor windrow composting operations, or totally enclosed in-vessel composting, sometimes referred to tunnel composting.*[2]

5.1 Aeration

The aeration system uses fans to push and/or pull air through the composting mass. Rigid or flexible perforated piping, connected to fans, delivers the air. The pipes can be installed in channels, on top of a floor, or included throughout the pile during buildup.

In large-scale systems, forced aeration is accompanied with a computerized monitoring system responsible for controlling the rate and schedule of air delivery to the composting mass, although meters and manual monitoring techniques may also be used in smaller scale operations.

Advantages of this composting method include the ability to maintain the proper moisture and oxygen levels for the microbial populations to operate at peak efficiency to reduce pathogens, while preventing excess heat, which can crash the system. Aerated systems also facilitate the use of biofilters to treat process air to remove particulates and mitigate odors prior to venting. However, aerated systems can dry out quickly and must be monitored closely to maintain desired moisture levels.

In Thailand this system has been used by 470 farmer groups. (May, 2008).*[3] The process required 30 days to finish without turning, with 10 metric tons of compost (10 piles) obtained each time. A 15-inch squirrel-cage blower was used to force the air through 10 static piles of compost, one at a time, for 15 minute periods twice a day. The raw materials consisted of agricultural wastes and animal manure in the ratio of 3:1 by volume.

5.2 See also

- List of composting systems

- Anaerobic digestion

Channeled concrete floor of a composting pad for perforated piping that delivers oxygen to the composting mass

- In-vessel composting
- List of solid waste treatment technologies
- Windrow composting

5.3 References

[1] Edmonton, AB, Canada Co-composting facility

[2] Batch tunnel composting in Europe

[3] Agricultural and Food Processing Wastes Composting In Thailand by Aerated Static Pile System

Aeration system for a closed chamber composting facility

Chapter 6

An Agricultural Testament

An Agricultural Testament is Sir Albert Howard's best-known publication, and remains one of the seminal works in the history of organic farming agricultural movement.*[1]*[2]*[3] Dedicated to his first wife and co-worker Gabrielle, herself a plant phsyiologist, it focuses on the nature and management of soil fertility, and notably explores composting.*[3] At a time when modern, chemical-based industrialized agriculture was just beginning to radically alter food production, it advocated natural processes rather than man-made inputs as the superior approach to farming. It was first published in England in 1940, with the first American edition in 1943.*[lower-alpha 1] Apart from a reprint by Rodale Press in 1972 and 1976 it remains out of print.*[4]

6.1 Notes

[1] The online version of *An Agricultural Testament* states that the first American edition was 1945. This is an error. The hardcopy of the 1976 Special Rodale Press Edition clearly states: "First American edition, 1943". The online text otherwise appears to be identical with the hardcopy, including the replication of the comments from the inside flaps and back of the dust-cover.

6.2 References

[1] Howard, Sir Albert (1943), *An Agricultural Testament* (PDF), Oxford, UK: Oxford University Press, retrieved 9 August 2010 pdf per Special Rodale Press Edition, 1976. See cover note on significance of book.

[2] William Lockeretz, ed. (2007), *Organic Farming: An International History*, Oxfordshire, UK & Cambridge, Massachusetts: CAB International (CABI), ISBN 978-0-85199-833-6, retrieved 10 August 2010 ebook ISBN 978-1-84593-289-3

[3] Michael Pollan (2006), *The Omnivore's Dilemma*, The Penguin Press, p. 145, ISBN 978-1-59420-082-3

[4] *An Agricultural Testament*, Out of print description, Cumberland Books website, retrieved 10 August 2010

6.3 External links

- *An Agricultural Testament* - full text online

Chapter 7

Biotic material

Biotic material or **biological derived material** is any material that originates from living organisms. Most such materials contain carbon and are capable of decay.

Examples of biotic materials are wood, linoleum, straw, humus, manure, bark, crude oil, cotton, spider silk, chitin, fibrin, and bone.

The use of biotic materials, and processed biotic materials (bio-based material) as alternative natural materials, over synthetics is popular with those who are environmentally conscious because such materials are usually biodegradable, renewable, and the processing is commonly understood and has minimal environmental impact. However, not all biotic materials are used in an environmentally friendly way, such as those that require high levels of processing, are harvested unsustainably, or are used to produce carbon emissions.

When the source of the recently living material has little importance to the product produced, such as in the production of biofuels, biotic material is simply called biomass. Many fuel sources may have biological sources, and may be divided roughly into fossil fuels, and biofuel.

In soil science, biotic material is often referred to as *organic matter*. Biotic materials in soil include glomalin, Dopplerite and humic acid. Some biotic material may not be considered to be organic matter if it is low in organic compounds, such as a clam's shell, which is an essential component of the living organism, but contains little organic carbon.

Examples of the use of biotic materials include:

- Alternative natural materials
- building material, for a stylistic reasons, or to reduce allergic reactions.
- clothing
- energy production
- food
- medicine
- ink
- composting and mulch

7.1 References

Chapter 8

Brown waste

Sawdust is an example of brown waste.

Brown Waste is any biodegradable waste that is predominantly carbon based. The term includes such items as grass cuttings, dry leaves, twigs, hay, paper, sawdust, corn cobs, cardboard, pine needles or cones, etc.*[1] Carbon is necessary to composting, which uses a combination of green waste and brown waste to promote the microbial processes involved in the decomposition process.*[2] The composting of brown waste sustainably returns the carbon to the carbon cycle.

8.1 See also

- Biomass

- Waste management

- Composting

8.2   References

[1] How to Make Compost, a Composting Guide

[2] California Integrated Waste Management Board - Home Composting

Chapter 9

Carbon-to-nitrogen ratio

The C/N ratio (C:N) or **carbon-to-nitrogen ratio** is a ratio of the mass of carbon to the mass of nitrogen in a substance. It can, amongst other things, be used in analysing sediments and compost. A useful application for C/N ratios is as a proxy for paleoclimate research, having different uses whether the sediment cores are terrestrial-based or marine-based. Carbon-to-nitrogen ratios are an indicator for nitrogen limitation of plants and other organisms and can identify whether molecules found in the sediment under study come from land-based or algal plants.[1] Further, they can distinguish between different land-based plants, depending on the type of photosynthesis they undergo. Therefore, the C/N ratio serves as a tool for understanding the sources of sedimentary organic matter, which can lead to information about the ecology, climate, and ocean circulation at different times in Earth's history.[2]

C/N ratios in the range 4-10:1 are usually from marine sources, whereas higher ratios are likely to come from a terrestrial source.[3] [4] Vascular plants from terrestrial sources tend to have C/N ratios greater than 20. [5] [6] The lack of cellulose, which has a chemical formula of $(C_6H_{10}O_5)_n$, and greater amount of proteins in algae versus vascular plants causes this significant difference in the C/N ratio. [7] [8] [9]

When composting, microbial activity utilizes a C/N ratio of 30-35:1 and a higher ratio will result in slower composting rates.[10] However, this assumes that carbon is completely consumed, which is often not the case. Thus, for practical agricultural purposes, a compost should have an initial C/N ratio of 20-30:1.[11]

Example of devices that can be used to measure this ratio are the CHN analyzer and the continuous-flow isotope ratio mass spectrometer (CF-IRMS). [12] However, for more practical applications, desired C/N ratios can be achieved by blending common used substrates of known C/N content, which are readily available and easy to use.

9.1 Applications

9.1.1 Marine

Organic matter that is deposited in marine sediments contains a key indicator as to its source and the processes it underwent before reaching the floor as well as after deposition, its carbon to nitrogen ratio.[13][14][15] In the global oceans, freshly produced algae in the surface ocean typically have a carbon to nitrogen ratio of about 4 to 10.[16] However, it has been observed that only 10% of this organic matter (algae) produced in the surface ocean sinks to the deep ocean without being degraded by bacteria in transit, and only about 1% is permanently buried in the sediment. An important process called sediment diagenesis accounts for the other 9% of organic carbon that sank to the deep ocean floor, but was not permanently buried, that is 9% of the total organic carbon produced is degraded in the deep ocean. [17] The microbial communities utilizing the sinking organic carbon as an energy source are partial to nitrogen-rich compounds because much of these bacterium are nitrogen-limited and much prefer it over carbon. As a result, the carbon to nitrogen ratio of sinking organic carbon in the deep ocean is elevated compared to fresh surface ocean organic matter that had not been degraded. An exponential increase in C/N ratios is observed with increasing water depth—with C/N ratios reaching 10 at intermediate water depths of about 1000 meters, and up to 15 in the deep ocean (~ >2500 meters).[18]

This elevated C/N signature is preserved in the sediment, until another form of diagenesis, post-depositional diagenesis, alters its C/N signature once again. Post-depositional diagenesis occurs in organic-carbon-poor marine sediments where bacteria are able to oxidize organic matter in aerobic conditions as an energy source. The oxidation reaction proceeds as follows: $CH_2O + H_2O \rightarrow CO_2 + 4H^*+ + 4e^*-$, with a standard free energy of -27.4 kJ mol[*]-1 (half reaction).[*][19] Once all of the oxygen is used up, bacteria are able to carry out an anoxic sequence of chemical reactions as an energy source, all with negative $\Delta G°r$ values, with the reaction becoming less favorable as the chain of reactions proceeds.[*][20]

The same principle described above explaining the preferential degradation of nitrogen-rich organic matter occurs within the sediments, as they are more labile and are in higher demand. This principle has been utilized in paleoceanographic studies in order to identify core sites that have not experienced much microbial activity, or contamination by terrestrial sources with much higher C/N ratios. [*][21]

Lastly, it should be noted that ammonia, the product of the second reduction reaction, which reduces nitrate and produces nitrogen gas and ammonia, is easily adsorbed on clay mineral surfaces and protected from bacteria. This has been proposed as an explanation for lower than expected C/N signatures of organic carbon in sediments that have undergone post-depositional diagenesis.[*][22]

Ammonium produced from the remineralisation of organic material, exists in elevated concentrations $(1 - >14\mu M)$ within cohesive shelf sea sediments found in the Celtic Sea (depth: 1-30cm). The depth of sediment exceeds 1m and would be a suitable study site to carry out paleolimnology experiments with C:N.

9.1.2   Lacustrine

Unlike in marine sediments, diagenesis does not pose a large threat to the integrity of the C/N ratio in lacustrine sediments.[*][23] [*][24]Though wood from living trees around lakes have consistently higher C/N ratios than wood buried in sediment, the change in elemental composition is not large enough to remove the vascular versus non-vascular plant signals due to the refractory nature of terrestrial organic matter.[*][25] [*][26] [*][27] Abrupt shifts in the C/N ratio down-core can be interpreted as shifts in the organic source material.

For example, two separate studies on Mangrove Lake, Bermuda and Lake Yunoko, Japan show irregular, abrupt fluctuations between C/N around 11 to around 18. These fluctuations are attributed to shifts from mainly algal dominance to land-based vascular dominance.[*][28][*][29] Results of studies that show abrupt shifts in algal dominance and vascular dominance often lead to conclusions about the state of the lake during these distinct periods of isotopic signatures. Times in which lakes are dominated by algal signals suggest the lake is a deep-water lake, while times in which lakes are dominated by vascular plant signals suggest the lake is shallow, dry, or marshy.[*][30] Using the C/N ratio in conjunction with other sediment observations, such as physical variations, D/H isotopic analyses of fatty acids and alkanes, and δ13C analyses on similar biomarkers can lead to further regional climate interpretations that describe the larger phenomena at play.

9.2   References

[1] Ishiwatari, R., and M. Uzaki. "Diagenetic Changes of Lignin Compounds in a More Than 0.6 Million-Year-Old Lacustrine Sediment (Lake Biwa, Japan)." Geochimica Et Cosmochimica Acta 51, no. 2 (Feb 1987): 321-28.

[2] Ishiwatari, R., and M. Uzaki. "Diagenetic Changes of Lignin Compounds in a More Than 0.6 Million-Year-Old Lacustrine Sediment (Lake Biwa, Japan)." Geochimica Et Cosmochimica Acta 51, no. 2 (Feb 1987): 321-28.

[3] Gray KR, Biddlestone AJ. 1973. Composting - process parameters. The Chemical Engineer. Feb. pp 71-76

[4] Stewart, Keith (2006). It's A Long Road to A Tomato. New York: Marlowe & Company. p. 155. ISBN 978-1-56924-330-5.

[5] Ishiwatari, R., and M. Uzaki. "Diagenetic Changes of Lignin Compounds in a More Than 0.6 Million-Year-Old Lacustrine Sediment (Lake Biwa, Japan)." Geochimica Et Cosmochimica Acta 51, no. 2 (Feb 1987): 321-28.

[6] Prahl, F. G., J. R. Ertel, M. A. Goni, M. A. Sparrow, and B. Eversmeyer. "Terrestrial Organic-Carbon Contributions to Sediments on the Washington Margin." Geochimica Et Cosmochimica Acta 58, no. 14 (Jul 1994): 3035-48.

[7] Ishiwatari, R., and M. Uzaki. "Diagenetic Changes of Lignin Compounds in a More Than 0.6 Million-Year-Old Lacustrine Sediment (Lake Biwa, Japan)." Geochimica Et Cosmochimica Acta 51, no. 2 (Feb 1987): 321-28.

[8] Meyers, Philip A., and Heidi Doose. "29. SOURCES, PRESERVATION, AND THERMAL MATURITY OF ORGANIC MATTER IN PLIOCENE–PLEISTOCENE ORGANIC-CARBON–RICH SEDIMENTS OF THE WESTERN MEDITER-RANEAN SEA." Proceedings of the Ocean Drilling Program: Scientific results. Vol. 161. The Program, 1999.

[9] Müller, P. J. "CN ratios in Pacific deep-sea sediments: Effect of inorganic ammonium and organic nitrogen compounds sorbed by clays." Geochimica et Cosmochimica Acta 41, no. 6 (1977): 765-776.

[10] Prahl, F. G., J. R. Ertel, M. A. Goni, M. A. Sparrow, and B. Eversmeyer. "Terrestrial Organic-Carbon Contributions to Sediments on the Washington Margin." Geochimica Et Cosmochimica Acta 58, no. 14 (Jul 1994): 3035-48.

[11] Dahlem. "Flux to the Seafloor", Group Report, eds. K.W. Bruland et al., pp. 210–213, 1988.

[12] Brenna, J. Thomas, et al. "High-precision continuous-flow isotope ratio mass spectrometry." Mass spectrometry reviews 16.5 (1997): 227-258.

[13] Jasper, J. P., and R. B. Gagosian. "The sources and deposition of organic matter in the Late Quaternary Pigmy Basin, Gulf of Mexico." Geochemica et Cosmochimica Acta 54, no. 4 (1990): 1117-1132.

[14] Meyers, P. A. "Preservation of Elemental and Isotopic Source Identification of Sedimentary Organic-Matter." Chemical Geology 114, no. 3-4 (Jun 1 1994): 289-302.

[15] Prahl, F. G., J. R. Ertel, M. A. Goni, M. A. Sparrow, and B. Eversmeyer. "Terrestrial Organic-Carbon Contributions to Sediments on the Washington Margin." Geochimica Et Cosmochimica Acta 58, no. 14 (Jul 1994): 3035-48.

[16] Meyers, P. A. "Preservation of Elemental and Isotopic Source Identification of Sedimentary Organic-Matter." Chemical Geology 114, no. 3-4 (Jun 1 1994): 289-302.

[17] Emerson, S., and J. Hedges. "Sediment Diagenesis and Benthic Flux." Treatise on Geochemistry 6.11 (2003): 293-319.

[18] Müller, P. J. "CN ratios in Pacific deep-sea sediments: Effect of inorganic ammonium and organic nitrogen compounds sorbed by clays." Geochimica et Cosmochimica Acta 41, no. 6 (1977): 765-776.

[19] Emerson, S., and J. Hedges. "Sediment Diagenesis and Benthic Flux." Treatise on Geochemistry 6.11 (2003): 293-319.

[20] Emerson, S., and J. Hedges. "Sediment Diagenesis and Benthic Flux." Treatise on Geochemistry 6.11 (2003): 293-319.

[21] Raymo, M. E., et al. "Mid-Pliocene warmth: stronger greenhouse and stronger conveyor." Marine Micropaleontology 27.1 (1996): 313-326.

[22] Müller, P. J. "CN ratios in Pacific deep-sea sediments: Effect of inorganic ammonium and organic nitrogen compounds sorbed by clays." Geochimica et Cosmochimica Acta 41, no. 6 (1977): 765-776.

[23] Ishiwatari, R., and M. Uzaki. "Diagenetic Changes of Lignin Compounds in a More Than 0.6 Million-Year-Old Lacustrine Sediment (Lake Biwa, Japan)." Geochimica Et Cosmochimica Acta 51, no. 2 (Feb 1987): 321-28.

[24] Meyers, Philip A., and Ryoshi Ishiwatari. "Lacustrine organic geochemistry—an overview of indicators of organic matter sources and diagenesis in lake sediments." Organic geochemistry 20.7 (1993): 867-900.

[25] Ishiwatari, R., and M. Uzaki. "Diagenetic Changes of Lignin Compounds in a More Than 0.6 Million-Year-Old Lacustrine Sediment (Lake Biwa, Japan)." Geochimica Et Cosmochimica Acta 51, no. 2 (Feb 1987): 321-28.

[26] Meyers, P. A. "Preservation of Elemental and Isotopic Source Identification of Sedimentary Organic-Matter." Chemical Geology 114, no. 3-4 (Jun 1 1994): 289-302.

[27] Meyers, Philip A., and Ryoshi Ishiwatari. "Lacustrine organic geochemistry—an overview of indicators of organic matter sources and diagenesis in lake sediments." Organic geochemistry 20.7 (1993): 867-900.

[28] Meyers, Philip A., and Ryoshi Ishiwatari. "Lacustrine organic geochemistry—an overview of indicators of organic matter sources and diagenesis in lake sediments." Organic geochemistry 20.7 (1993): 867-900.

[29] Ishiwatari, R., N. Takamatsu, and T. Ishibashi. "Separation of autochthonous and allochthonous materials in lacustrine sediments by density differences."Japanese Journal of Limnology 38 (1977).

[30] Meyers, Philip A., and Ryoshi Ishiwatari. "Lacustrine organic geochemistry—an overview of indicators of organic matter sources and diagenesis in lake sediments." Organic geochemistry 20.7 (1993): 867-900.

9.3   External links

- Compost and C/N Ratio

- C/N calculator

Chapter 10

Composting Association

The **Organics Recycling Group** (ORG), formerly the Association for Organics Recycling (AfOR) and before that the Composting Association, is the leading trade organisation for the biodegradable waste management industry in the UK. It helped to develop the BSI PAS 100 industry standard for composts.

ORG was formed by the merger of AfOR and the Renewable Energy Association (REA) on 1 January 2013 which created a membership of around 1,100 companies, organisations and individuals.[*][1]

To main objective of the group is to promote the sustainable management of biodegradable resources, covering both aerobic and anaerobic technologies such as windrow and in-vessel composting, anaerobic digestion and mechanical biological treatment. ORG specialises in issues covering the collection, treatment and use of these resources, to complement the work undertaken by the REA on matters such as distributed generation, financial incentives, the Renewables Obligation and planning.

10.1   See also

- Anaerobic digestion
- BSI PAS 100
- Compost
- Composting

10.2   References

[1]   "Organics Recycling Group's terms of reference, aims and objectives" . *organics-recycling.org.uk*. Retrieved 2014-10-02.

10.3   External links

- Organics Recycling Group website

Chapter 11

Composting toilet

A **composting toilet** is a type of dry toilet that uses a predominantly aerobic processing system to treat human excreta, by composting or managed aerobic decomposition. These toilets generally use little to no water and may be used as an alternative to flush toilets.*[1] They have found use in situations where no suitable water supply or sewer system and sewage treatment plant is available to capture the nutrients in human excreta. They are in use in many roadside facilities and national parks in Sweden, Canada, US, UK and Australia. They are used in rural holiday homes in Sweden and Finland.

The human excreta is usually mixed with sawdust, coconut coir or peat moss to facilitate aerobic processing, liquid absorption, and odor mitigation. Most composting toilets use slow, cold composting conditions, sometimes connected to a secondary external composting step.

Composting toilets produce a compost that may be used for horticultural or agricultural soil enrichment if the local regulations allow this. A curing stage is often needed to allow mesophilic composting to reduce potential phytotoxins.

11.1   Terminology

The term "composting toilet" is used quite loosely, and its meaning may vary by country. For example, in English-speaking countries, the term "anaerobic composting" (equivalent to anaerobic decomposition) is used. In Germany and Scandinavian countries, composting always refers to a predominantly aerobic process. This aerobic composting may take place with an increase in temperature due to microbial action, or without a temperature increase in the case of slow composting or cold composting. If earth worms are used (vermicomposting) then there is also no increase in temperature.

Composting toilets differ from pit latrines, arborloo or tree bogs, all of which are forms of less controlled decomposition and may not protect groundwater from nutrient or pathogen contamination or provide optimal nutrient recycling. They also differ from urine-diverting dry toilets (UDDTs) where pathogen reduction is achieved through dehydration (also known by the more precise term "desiccation") and where the faeces collection vault is kept as dry as possible. Composting toilets target a certain degree of moisture in the composting chamber.

Composting toilets usually do not divert urine. Offering a waterless urinal in addition to the toilet can help keep excess amounts of urine out of the composting chamber.

Composting toilets can be used to implement an ecological sanitation (ecosan) approach for resource recovery. However, ecosan is an approach and not a specific technology. Therefore, these toilets should not be called "ecosan toilets".

Composting toilets have also been called "sawdust toilets", which can be appropriate if the amount of aerobic composting taking place in the toilet's container is very limited.*[2] The "Clivus multrum" is a type of composting toilet which has a large composting chamber below the toilet seat and also receives undigested organic material to increase the carbon to nitrogen ratio.

Composting toilet at Activism Festival 2010 in the mountains outside Jerusalem

11.2 Applications

Composting toilets can be suitable in areas such as a rural area or a park that lacks a suitable water supply, sewers and sewage treatment. They can also be help to increase the resilience of existing sanitation systems in the face of possible natural disasters such climate change, earthquakes or tsunami. Composting toilets can reduce or perhaps eliminate the need for a septic tank system to reduce environmental footprint (particularly when used in conjunction with an on-site

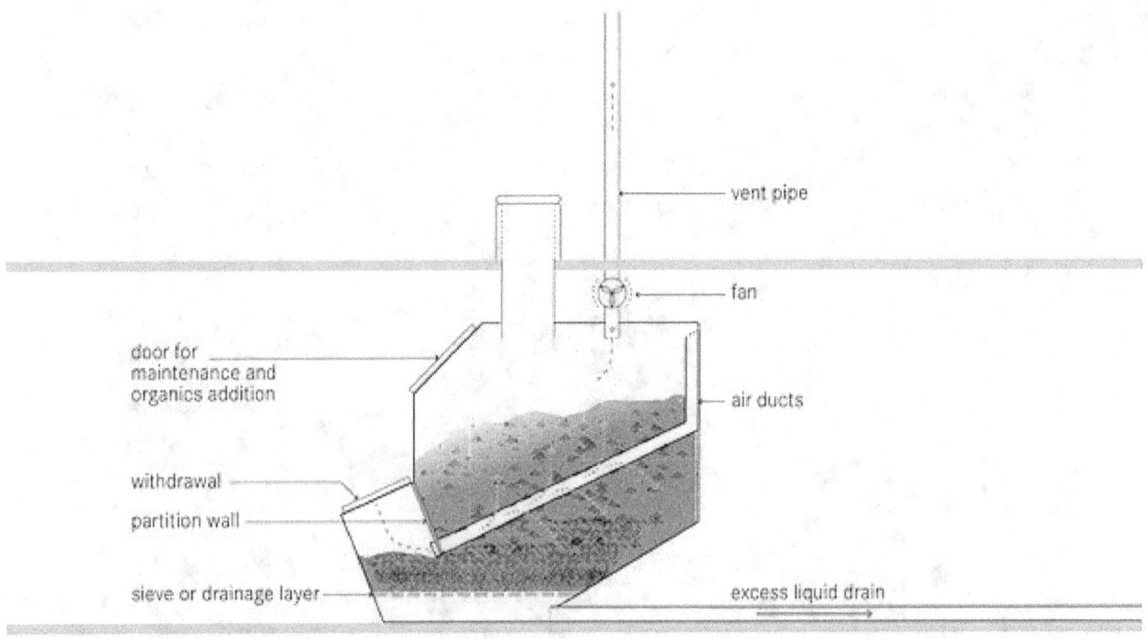

Schematic of the composting chamber which is located below the toilet seat[1]

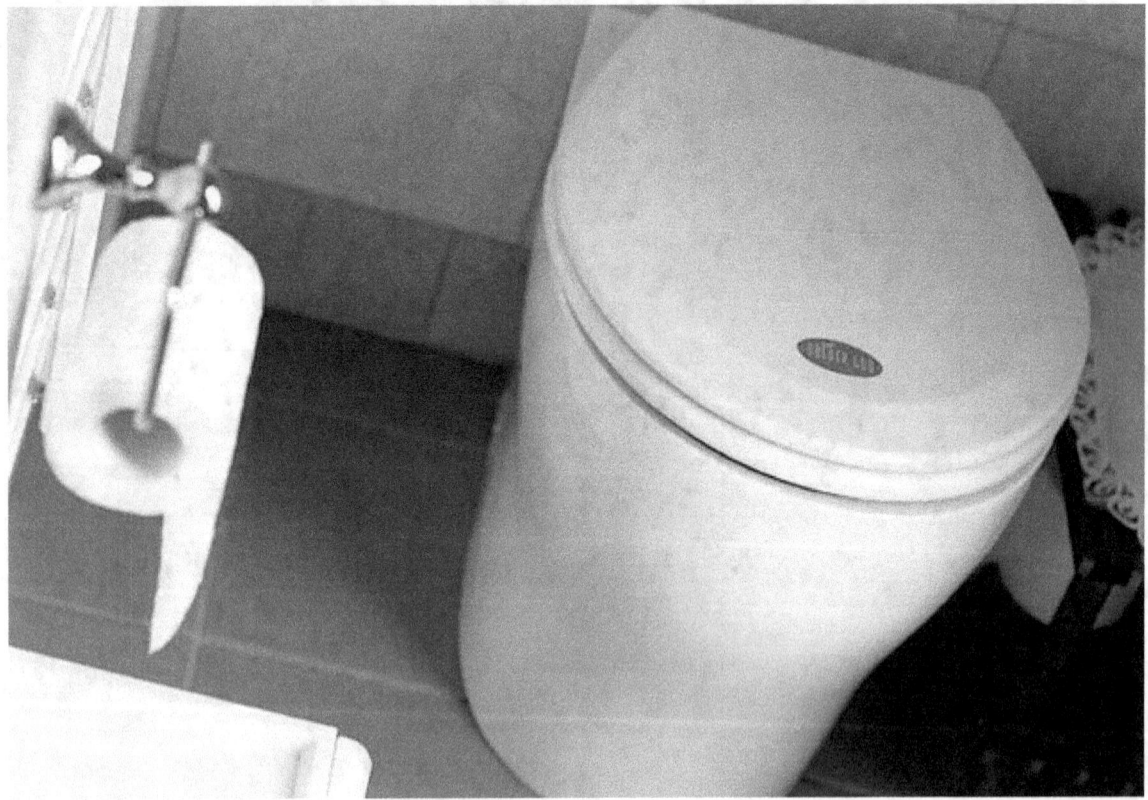

This is the pedestal for a split-system composting toilet where collection/treatment chambers are located below the bathroom floor.

Inexpensive do-it-yourself compost toilet at Dial House, Essex, England, utilizing an old desk as the toilet unit.

greywater treatment system).

These types of toilets can be used for resource recovery by reusing sanitized feces and urine as fertilizer and soil conditioner for gardening or ornamental activities.

11.3 Basics

Main article: Compost

11.3.1 Components

A composting toilet consists of two elements: a place to sit or squat and a collection/composting unit.*[3] The composting unit consists of four main parts:*[1]

- storage or composting chamber

- a ventilation unit to ensure that the degradation process in the toilet is predominantly aerobic and to vent odorous gases

- a leachate collection system to remove excess liquid

- an access door for extracting the compost

Public composting toilet at a highway rest facility in Sweden

11.3.2   Construction

The composting chamber can be constructed above or below ground level. It can be inside a structure or include a separate superstructure.

A drainage system removes leachate. Otherwise, excess moisture can cause anaerobic conditions and impede decomposition. Urine diversion can improve compost quality, since urine contains a large amounts of ammonia that inhibits microbiological activity.*[1]

Composting toilets greatly reduce human waste (excreta) volumes through psychrophilic, thermophilic or mesophilic composting. Keeping the composting chamber insulated and warm protects the composting process from slowing due to low temperatures.

11.3.3   Odorous gases

The following gases may be emitted during the composting process that takes place in composting toilets: hydrogen sulfide (H_2S), ammonia, nitrous oxide (N_2O) and volatile organic compounds (VOCs).*[4] These gases can potentially lead to complaints about odours. Some methane may also be present, but it is not odorous.

11.4 Pathogen removal

Excreta-derived compost recycles fecal nutrients, but it can carry and spread pathogens if the process of reuse of excreta is not done properly.

Internal pathogen destruction rates are usually low, particularly helminth eggs, such as Ascaris eggs,*[2] risking disease absent proper system management. Compost from human excreta processed under only mesophilic conditions or taken directly from the compost chamber is not safe for food production.*[5] High temperatures or long composting times are required to kill helminth eggs, the hardiest of all pathogens. Helminth infections are common in many developing countries.

In thermophilic composting bacteria that thrive at temperatures of 40–60 °C (104–140 °F) oxidize (break down) waste into its components, some of which are consumed in the process, reducing volume and eliminating potential pathogens. To destroy pathogens thermophilic composting must heat the compost pile sufficiently, or enough time (1–2 years) must elapse since fresh material was added that biological activity has had the same pathogen removal effect.

One guideline claims that pathogen levels are reduced to a safe level by thermophilic composting at temperatures of 55 °C for at least two weeks or at 60 °C for one week.*[3] An alternate guideline claims that complete pathogen destruction may be achieved already if the entire compost heap reaches a temperature of 62 °C (144 °F) for one hour, 50 °C (122 °F) for one day, 46 °C (115 °F) for one week or 43 °C (109 °F) for one month,*[1] although others regard this as overly optimistic.*[3]

11.5 Design considerations

11.5.1 Environmental factors

Four main factors affect the decomposition process:*[1]

- Sufficient oxygen is necessary for aerobic composting

- Moisture content from 45 to 70 percent (heuristically, "the compost should feel damp to the touch, with only a drop or two of water expelled when tightly squeezed in the hand." *[3])

- Temperature between 40 to 50 °C (achieved through proper chamber dimensioning and possibly active mixing)

- Carbon-to-nitrogen ratio (C:N) of 25:1

11.5.2 Additives and bulking material

Human excreta and food waste do not provide optimum conditions for composting. Usually the water and nitrogen content is too high, particularly when urine is mixed with feces. Additives or "bulking material", such as wood chips, bark chips, sawdust, ash and pieces of paper can absorb moisture. The additives improve pile aeration and increase the carbon to nitrogen ratio.*[3] Bulking material also covers faeces and reduces insect access. Absent sufficient bulking material, the material may become too compact and form impermeable layers, which leads to anaerobic conditions and odour.*[3]

11.5.3 Leachate management

Leachate removal controls moisture levels, which is necessary to ensure rapid, aerobic composting. Some commercial units include a urine-separator or urine-diverting system and/or a drain at the bottom of the composter for this purpose.

11.5.4   Aeration and mixing

Microbial action also requires oxygen, typically from the air. Commercial systems provide ventilation that moves air from the bathroom, through the waste container, and out a vertical pipe, venting above the roof. This air movement (via convection or fan forced) passes carbon dioxide and odors.

Some units require manual methods for periodic aeration of the solid mass such as rotating the composting chamber or pulling an "aerator rake" through the mass.

11.6   Types

Commercial units and construct-it-yourself systems are available.*[6] Variations include number of composting vaults, removable vault, urine diversion and active mixing/aeration.*[3]

11.6.1   Slow composting (or moldering) toilets

Most composting toilets use slow composting which is also called "cold composting". The compost heap is built up step by step over time.

The finished end product from "slow" composting toilets ("moldering toilets" or "moldering privies" in the US), is generally not free of pathogens. World Health Organization Guidelines from 2006 offer a framework for safe reuse of excreta, using a multiple barrier approach.*[7]

Slow composting toilets employ a passive approach. Common applications involve modest and often seasonal use, such as remote trail networks. They are typically designed such that the materials deposited can be isolated from the operational part. The toilet can also be closed to allow further mesophilic composting.*[7] Slow composting toilets rely on long retention times for pathogen reduction and for decomposition of excreta or on the combination of time and/or the addition of red wriggler worms for vermi-composting. Worms can be introduced to accelerate composting. Some jurisdictions of the US consider these worms as invasive species.*[7]

Example in Vermont woods

Slow composting toilets have been installed by the Green Mountain Club in Vermont's woodlands. They employ multiple vaults (called cribs) and a movable building. When one of the vaults fills, the building is moved over an empty vault. The full vault is left untouched for as long as possible (up to three years) before it is emptied. The large surface area and exposure to air currents can cause the pile to dry out. To counteract this, signs instruct users to urinate in the toilet.*[8] The club also uses pit latrines and simple bucket toilets with woodchips and external composting and directs users to urinate in the forest to prevent odiferous anaerobic conditions.*[9]

11.6.2   Active composters

Self-contained

"Self-contained" composting toilets compost in a container within the toilet unit. They are slightly larger than a flush toilet, but use roughly the same floor space. Some units use fans for aeration, and optionally, heating elements to maintain optimum temperatures to hasten the composting process and to evaporate urine and other moisture. Operators of composting toilets commonly add a small amount of absorbent carbon material (such as untreated sawdust, coconut coir, peat moss) after each use to create air pockets to encourage aerobic processing, to absorb liquid and to create an odor barrier. This additive is sometimes referred to as "bulking agent." Some owner-operators use microbial "starter" cultures to ensure composting bacteria are in the process, although this is not critical.

Remote

"Remote" "central" or "underfloor" units collect excreta via a toilet stool, either waterless, vacuum or micro-flush, from which it drains into a composter. "Vacuum-flush systems" can flush horizontally or upward with a small amount of water to the composter. "Micro-flush" toilets use about 500 millilitres (17 US fl oz) per use. These units feature a chamber below the toilet stool (such as in a basement or outside) where composting takes place and are suitable for high-volume and year-round applications as well as to serve multiple toilet stools.*[10]

11.6.3 Other

Some units employ roll-away containers fitted with aerators, while others use sloped-bottom tanks.

11.7 Maintenance

Maintenance is critical to ensure proper operation, including odor prevention. Maintenance tasks include: cleaning, servicing technical components such as fans and removal of compost, leachate and urine. Urine removal is only required for those types of composting toilets using urine diversion.

Once composting is complete (or more often), the compost must be removed from the unit. How often this occurs is a function of container size, usage and composting conditions, such as temperature.*[3] Active, hot composting may span months only while passive, cold composting may require years. Properly managed units yield output volumes of about 10% of inputs.

11.8 Uses of compost

Main articles: Uses of compost and Reuse of excreta

The material from composting toilets is a humus-like material, which can be suitable as a soil amendment for agriculture. Compost from residential composting toilets can be used in domestic gardens, and this is the main such use.

Enriching soil with compost adds substantial nitrogen, phosphorus, potassium, carbon and calcium. In this regard compost is equivalent to many fertilizers and manures purchased in garden stores. Compost from composting toilets has a higher nutrient availability than the dried faeces that result from a urine-diverting dry toilet.*[3]

Urine is typically present, although some is lost via leaching and evaporation. Urine can contain up to 90 percent of the residual nitrogen, up to 50 percent of the phosphorus, and up to 70 percent of the potassium.*[11]

Compost derived from these toilets has in principle the same uses as compost derived from other organic waste products, such as sewage sludge or municipal organic waste. However, users of excreta-derived compost must consider the risk of pathogens.

11.8.1 Pharmaceutical residues

Excreta-derived compost may contain prescription pharmaceuticals. Such residues are also present in conventional wastewater treatment effluent. This could contaminate groundwater. Among the medications that have been found in groundwater in recent years are antibiotics, antidepressants, blood thinners, ACE inhibitors, calcium-channel blockers, digoxin, estrogen, progesterone, testosterone, Ibuprofen, caffeine, carbamazepine, fibrates and cholesterol-reducing medications.*[12] Between 30% and 95% of pharmaceuticals medications are excreted by the human body. Medications that are lipophilic (dissolved in fats) are more likely to reach groundwater by leaching from fecal wastes. Wastewater treatment plants remove an average of 60% of these medications.*[13] The percentage of medications degraded during composting of excreta has not yet been reported.

11.9   Comparison

11.9.1   Pit latrines

Unlike pit latrines, composting toilets convert feces into a dry, odorless material, avoiding the issues surrounding liquid fecal sludge management (e.g. odor, insects and disposal). These toilets minimize the risk of water pollution through the safe containment of feces in above-ground vaults, which allows the toilets to be sited in locations where pit-based systems are not appropriate.

However, composting toilets face higher capital costs (although lifecycle costs might be lower) and greater complexity (for instance, adding covering materials and managing moisture content).

11.9.2   Flush toilets

Unlike flush toilets, composting toilets do not dilute excreta and create wastewater streams which must be treated before disposal. On the other hand, wastewater treatment plants can centralize waste management for an entire community, with potentially greater efficiency.

11.9.3   Urine-diverting dry toilets

Composting toilets are more difficult to maintain than other types of dry toilets, like urine-diverting dry toilets (UDDT) with which they are often confused. This is due to the need to maintain a consistent and relatively high moisture content. Apart from that, composting toilets are quite similar to UDDTs, sharing many of the same advantages and disadvantages.

11.10   History

11.10.1   Dry earth toilet

Before the flush toilet became accepted in the late 19th century in developed countries, some inventors, scientists and public health officials supported the use of "dry earth closets", a type of dry toilet with similarities to composting toilets. However, the collection vessel for the human excreta was not designed to compost. Dry earth closets were invented by English clergyman Henry Moule, who dedicated his life to improving public sanitation after witnessing the cholera epidemics of 1849 and 1854. Impressed by the insalubrity of the houses, especially during the Great Stink in the summer of 1858, he invented what he called the 'dry earth system'.

In partnership with James Bannehr, he patented his device (No. 1316, dated 28 May 1860). Among his works bearing on the subject were *The Advantages of the Dry Earth System* (1868), *The Impossibility overcome: or the Inoffensive, Safe, and Economical Disposal of the Refuse of Towns and Villages* (1870), The *Dry Earth System* (1871), *Town Refuse, the Remedy for Local Taxation* (1872), and *National Health and Wealth promoted by the general adoption of the Dry Earth System* (1873).

His system was adopted in private houses, in rural districts, in military camps, in many hospitals, and extensively in the British Raj. Ultimately, however, it failed to gain public support as attention turned to the water-flushed toilet connected to a sewer system.

In Germany, a similar dry toilet with a peat dispenser was marketed until after the second World War (it was called "Metroclo" and was manufactured by Gefinal, Berlin).

11.11   Society and culture

11.11.1 Regulations

International Organization for Standardization (ISO)

The International Organization for Standardization (ISO) is currently preparing a "management standard". As of 2015 this was in a draft state as ISO 24521, under the heading "Activities relating to drinking water and wastewater services — Guidelines for the management of basic onsite domestic wastewater services".[14] The standard is meant to be used in conjunction with ISO 24511.[15] It deals with toilets (including composting toilets) and toilet waste. The guidelines are applicable to basic wastewater systems and include the complete domestic wastewater cycle, such as planning, usability, operation and maintenance, disposal, reuse and health.

International Association of Plumbing and Mechanical Officials

The International Association of Plumbing and Mechanical Officials (IAPMO) is a plumbing and mechanical code structure adopted by many developed countries. It recently proposed an addition to its "Green Plumbing Mechanical Code Supplement" that, "...outlines performance criteria for site built composting toilets with and without urine diversion and manufactured composting toilets."[16] If adopted, this composting and urine diversion toilet code (the first of its kind in the United States) will appear in the 2015 edition of the Green Supplement to the Uniform Plumbing Code.[17][18]

United States

No performance standards for composting toilets are universally accepted in the US. Seven jurisdictions in North America[19] use *American National Standard/NSF International Standard ANSI/NSF 41-1998: Non-Liquid Saturated Treatment Systems*. An updated version was published in 2011.[20][note 1] Systems might also be listed with the Canadian Standards Association, cETL-US, and other standards programs.

Regarding byproduct regulation, several US states permit disposal of solids from composting toilets (usually a distinction between different types of dry toilets is not made) by burial, with varying or no minimum depth mandates (as little as 6 inches). For instance:

- Massachusetts: "Residuals from the composting toilet system must be buried on-site and covered with a minimum of six inches of clean compacted soil.[21] Massachusetts requires that any liquids produced but, "not recycled through the toilet [itself be] either discharged through a greywater system on the property that includes a septic tank and soil absorption system, or removed by a licensed septage hauler."[21]

- Oregon: "Humus from composting toilets may be used around ornamental shrubs, flowers, trees, or fruit trees and shall be buried under at least twelve inches of soil cover."[22]

- Rhode Island: "Solids produced by alternative toilets may be buried on site," while, "residuals shall not be applied to food crops."[23]

- Virginia: "All materials removed from a composting privy shall be buried," and "compost material shall not be placed in vegetable gardens or on the ground surface."[24]

- Vermont: "Byproducts may be disposed via "...shallow burial in a location approved by the Agency that meets the minimum site conditions [required for an onsite septic tank-based sanitation system]."[25]

- Washington: models its extensive regulations for what it refers to as "waterless toilets" on the federal regulations that govern sewage sludge.[26]

The Environmental Protection Agency has no jurisdiction over the byproducts of a dry toilet as long as excreta are not referred to as "fertilizer" (but instead simply a material that is being disposed of). Federal rule 503, known colloquially as the "EPA Biosolids rule" or the "EPA sludge rule" applies only to fertilizer. Thus, individual states regulate composting toilets.[27][28]

Germany

The regulations for composting toilets and other forms of dry toilets in Germany vary from state to state and from one application to another (e.g. use in allotment gardens or use in family homes and settlements). In the different states of Germany, it is the "Landesbauordnung" (translates to "state civil engineering regulations") of the respective state that regulates the use of such alternative toilets.[*][29] Most of them stipulate the use of flush toilets, however there are many exceptions, for example in the states of Hamburg, Lower Saxony, Bavaria, Mecklenburg-Western Pomerania, Rhineland-Palatinate, Saxony-Anhalt and Thuringia.[*][29] These generally make exceptions for the use of composting toilets in homes provided that there are no concerns for public health.

Regulations governing the use of compost and urine from composting toilets is less clear in Germany but it seems generally allowed provided it is used on one's own property and not sold to third parties.[*][29]

11.11.2 Examples

Finland

Numerous sparsely settled villages in rural areas in Finland are not connected to municipal water supply or sewer networks, requiring homeowners to operate their own systems. Individual private wells, i.e. shallow dug wells or boreholes in the bedrock, are often used for water supply, and many homeowners have opted for composting toilets. In addition, these toilets are common at holiday homes, often located near sensitive water bodies. For these reasons, many manufacturers of composting toilets are based in Finland, including Biolan, Ekolet, Kekkilä, Pikkuvihreä and Raita Environment.[*][30][*][31]

Estimates made by leading Finnish composting toilet manufacturers and the Global Dry Toilet Association of Finland provided the following 2014 figures for composting toilet use in Finland:

- About 4% of single-family homes not connected to a public sewer network are equipped with a composting toilet.

- Some 200,000 manufactured composting toilets are thought to serve holiday homes, matched by the number of other dry toilets. The simplest ones are sited in an outhouse.

Germany

Composting toilets have been successfully installed in houses with up to four floors.[*][3] An estimate from 2008 put the number of composting toilets in households in Germany at 500.[*][32] Most of these residences are also connected to a sewer system; the composting toilet was not installed due a lack of sewer system but for other reasons, mainly because of an "ecological mindset" of the owners.

In Germany and Austria, composting toilets and other types of dry toilets have been installed in single and multi-family houses (e.g. Hamburg, Freiburg, Berlin), ecological settlements (e.g. Hamburg-Allemöhe, Hamburg-Braamwisch, Kiel-Hassee, Bielefeld-Waldquelle, Wien-Gänserndorf) and in public buildings (e.g. Ökohaus Rostock, VHS-Ökostation Stuttgart-Wartberg, public toilets in recreational areas, restaurants and huts in the Alps, house boats and forest Kindergartens).[*][32]

The ecological settlement in Hamburg-Allermöhe has had composting toilets since 1982. The settlement of 36 single-family houses with approximately 140 inhabitants uses composting toilets, rainwater harvesting and constructed wetlands. Composting toilets save about 40 litres of water per capita per day compared to a conventional flush toilet (10 liter per flush), which adds up to 2,044 m^3 water savings per year for the whole settlement.[*][33]

Worldwide

Composting toilets with a large composting container (of the type Clivus Multrum and derivations of it) are popular in US, Canada, Australia, New Zealand and Sweden. They can be bought and installed as commercial products, as designs for self builders or as "design derivatives" which are marketed under various names. It has been estimated that approximately 10,000 such toilets might be in use worldwide.

11.12 Notes

[1] A listing of the most current NSF/ANSI standards can be found in PDF format at NSF International's *Standards* subdomain.

11.13 References

[1] Tilley, E., Ulrich, L., Lüthi, C., Reymond, Ph., Zurbrügg, C. *Compendium of Sanitation Systems and Technologies - (2nd Revised Edition)*. Swiss Federal Institute of Aquatic Science and Technology (Eawag), Duebendorf, Switzerland. ISBN 978-3-906484-57-0.

[2] Hill, B. G. (2013). An evaluation of waterless human waste management systems at North American public remote sites. PhD thesis, University of British Columbia (Vancouver), Canada

[3] Berger, W. (2011). Technology review of composting toilets - Basic overview of composting toilets (with or without urine diversion). Deutsche Gesellschaft für Internationale Zusammenarbeit (GIZ) GmbH, Eschborn, Germany

[4] Font, Xavier; Artola, Adriana; Sánchez, Antoni (6 April 2011). "Detection, Composition and Treatment of Volatile Organic Compounds from Waste Treatment Plants". *Sensors* **11** (12): 4043–4059. doi:10.3390/s110404043.

[5] Stenström, T.A., Seidu, R., Ekane, N., Zurbrügg, C. (2011). Microbial exposure and health assessments in sanitation technologies and systems - EcoSanRes Series, 2011-1. Stockholm Environment Institute (SEI), Stockholm, Sweden, page 88

[6] National Small Flows Clearinghouse, West Virginia University, Composting toilet technology

[7] WHO (2006). WHO Guidelines for the Safe Use of Wastewater, Excreta and Greywater - Volume IV: Excreta and greywater use in agriculture. World Health Organization (WHO), Geneva, Switzerland

[8] Allen, Lee (2013). "Long Trail News: Quarterly of the Green Mountain Club, Fall 2013. Article titled: "A Privy is a Privy is a Privy...or is it? To Pee or Not Pee."" (PDF). *Green Mountain Club*. Green Mountain Club. Retrieved 31 January 2013.

[9] Antos-Ketcham, Pete (2013). "Long Trail News: Quarterly of the Green Mountain Club, Fall 2013. Article titled: "Batch-Bin/Beyond-the-Bin (BTB) Composting Privies"" (PDF). *Green Mountain Club*. Green Mountain Club. Retrieved 31 January 2015.

[10] Berger, W. (2009). Appendix of technology review of composting toilets - List of manufacturers and commercially available composting toilets. Gesellschaft für Internationale Zusammenarbeit (GIZ) GmbH

[11] J.O. Drangert, Urine separation systems

[12] *Drugs in the Water*. Harvard Health Letter. 2011.

[13] Encyclopedia of Quantitative Risk Analysis and Assessment, Volume 1, edited by Edward L. Melnick, Brian S. Veritt, 2008

[14] "ISO/DIS 24521. Activities relating to drinking water and wastewater services -- Guidelines for the management of basic onsite domestic wastewater services". *International Organization for Standardization (ISO)*. Retrieved 15 January 2015.

[15] "ISO 24511:2007. Activities relating to drinking water and wastewater services -- Guidelines for the management of wastewater utilities and for the assessment of wastewater services". *International Organization for Standardization (ISO)*. Retrieved 15 January 2015.

[16] "Recode September 2014 Newsletter". *Recode*. Recode. September 2014. Retrieved 15 January 2015.

[17] "IAPMO Proposed Composting and Urine DIversion Toilet Code" (PDF). *The IAPMO Group*. International Association of Plumbing and Mechanical Officials. Retrieved 15 January 2015.

[18] Cole, Daniel (January 2015). "IAPMO GPMCS raising the bar for water, energy efficiency". *Plumbing Engineer*. Plumbing Engineer. Retrieved 15 January 2015.

[19] Oregon Onsite Advisory Committee "Final Report of Recommended Changes to Rules Governing Onsite Systems", *OR DEQ*, February 8, 2010, accessed May 8, 2011.

[20] "PUBLICATIONS - Standards and Criteria - March 21, 2013" (PDF). NSF International. p. 4. Retrieved 24 March 2013. Wastewater Treatment Units ···NSF/ANSI 41 – 2011: Non-liquid saturated treatment systems (composting toilets)

[21] "Regulatory Provisions for Composting Toilets and Greywater Systems". *The Official Website of the Massachusetts Executive Office of Energy and Environmental Affairs*. Office of Energy and Environmental Affairs. Retrieved 13 January 2015.

[22] "Department of Consumer and Business Services, Building Codes Division, Division 770, Plumbing Product Approvals". *Oregon Secretary of State*. State of Oregon. Retrieved 13 January 2015.

[23] "State of Rhode Island and Providence Plantations Department of Environmental Management, Office of Water Resources: "Rules Establishing Minimum Standards Relating to Location, Design, Construction and Maintenance of Onsite Wastewater Treatment Systems"" (PDF). *State of Rhode Island Department of Environmental Management*. STATE OF RHODE ISLAND AND PROVIDENCE PLANTATIONS. July 2010. Retrieved 13 January 2015.

[24] "SEWAGE HANDLING AND DISPOSAL REGULATIONS (Emergency Regulations for Gravelless Material and Drip Dispersal), 12 VAC 5-610-10 et seq." (PDF). *State of Virginia Department of Health*. Commonwealth of Virginia. 14 March 2014. Retrieved 13 January 2015.

[25] "Environmental Protection Rules, Chapter 1: Wastewater System and Potable Water Supply Rules" (PDF). *State of Vermont Drinking Water and Groundwater Protection Division*. State of Vermont. 29 September 2007. Retrieved 14 January 2015.

[26] "Recommended Standards and Guidance for Performance, Application, Design, and Operation & Maintenance: Water Conserving On-Site Wastewater Treatment Systems" (PDF). *State of Washington Department of Health*. State of Washington. July 2012. Retrieved 14 January 2015.

[27] "Water Efficiency Technology Fact Sheet: Composting Toilets" (PDF). *United States Environmental Protection Agency, Office of Water, Washington, D.C., EPA 832-F-99-066*. United States Environmental Protection Agency, Office of Water. September 1999. Retrieved 13 January 2015.

[28] "TITLE 40—Protection of Environment, Chapter I—Environmental Protection Agency (Continued), Subchapter O—Sewage Sludge, Part 503—Standards for the Use or Disposal of Sewage Sludge". *Electronic Code of Federal Regulations*. United States Government Publishing Office. Retrieved 13 January 2015.

[29] Lorenz-Ladener, Hrsg. Claudia; Berger, Wolfgang (2005). *Kompost-Toiletten: Wege zur sinnvollen Fäkalienentsorgung* (1. überarb. u. erw. Aufl. ed.). Staufen im Breisgau: Ökobuch. p. 178. ISBN 978-3-936896-16-9.

[30] Global Dry Toilet Association of Finland (2011) Dry Toilet Manufacturers in Finland, Leaflet in English and Finnish

[31] "Global Dry Toilet Association of Finland". *Global Dry Toilet Association of Finland - Company and association members*. Retrieved 15 January 2015.

[32] Lorenz-Ladener, Hrsg. Claudia; Berger, Wolfgang (2005). *Kompost-Toiletten: Wege zur sinnvollen Fäkalienentsorgung* (1. überarb. u. erw. Aufl. ed.). Staufen im Breisgau: Ökobuch. p. 183. ISBN 978-3-936896-16-9.

[33] Rauschning, G., Berger, W., Ebeling, B., Schöpe, A. (2009). Ecological settlement in Allermöhe Hamburg, Germany - Case study of sustainable sanitation projects. Sustainable Sanitation Alliance (SuSanA)

11.14   External links

- "What is a Composting Toilet System and How Does it Compost?"

- Composting toilet description (Sustainable Sanitation and Water Management Toolbox)

- Composting systems (documents in library of Sustainable Sanitation Alliance)

- More photos of composting toilets in Flickr photo database of Sustainable Sanitation Alliance

Composting toilet with a seal in the lid in Germany

External composting chamber of a composting toilet at a house in France

Finished compost from a composting toilet ready for application as soil improvement in Kiel-Hassee, Germany

Henry Moule's earth closet, patented in 1873 (not a true composting toilet). Example from around 1875. Rear chamber for dispensing cover material

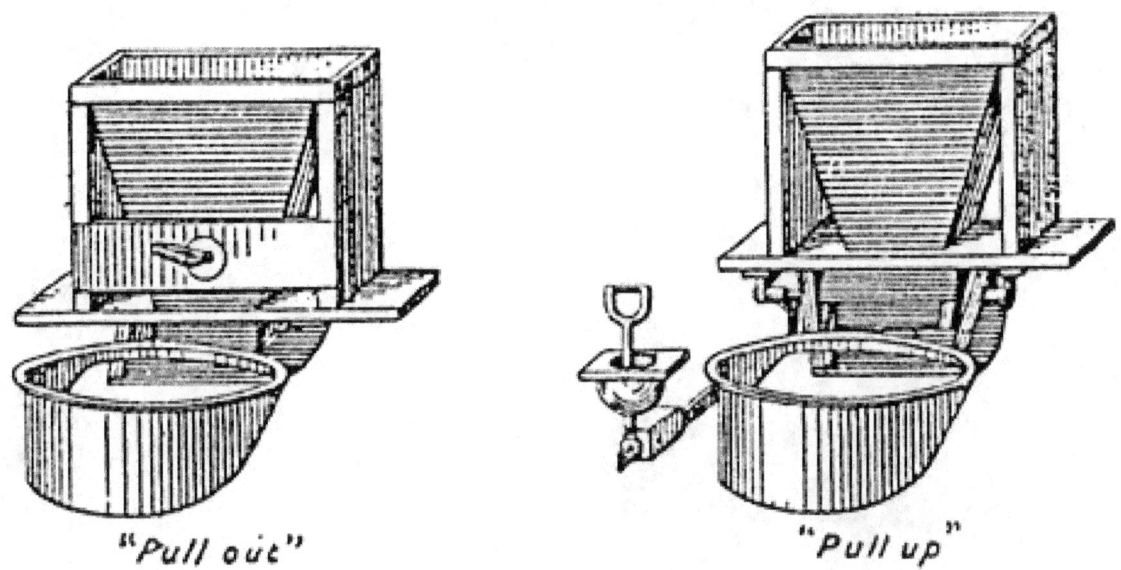

"Pull out"
"Pull up"

Figs. 8 and 9.—Moule's Earth Closets.

Section
8½
1'7"
1'6"
1'5"
1'5¼"
1'11"
10"

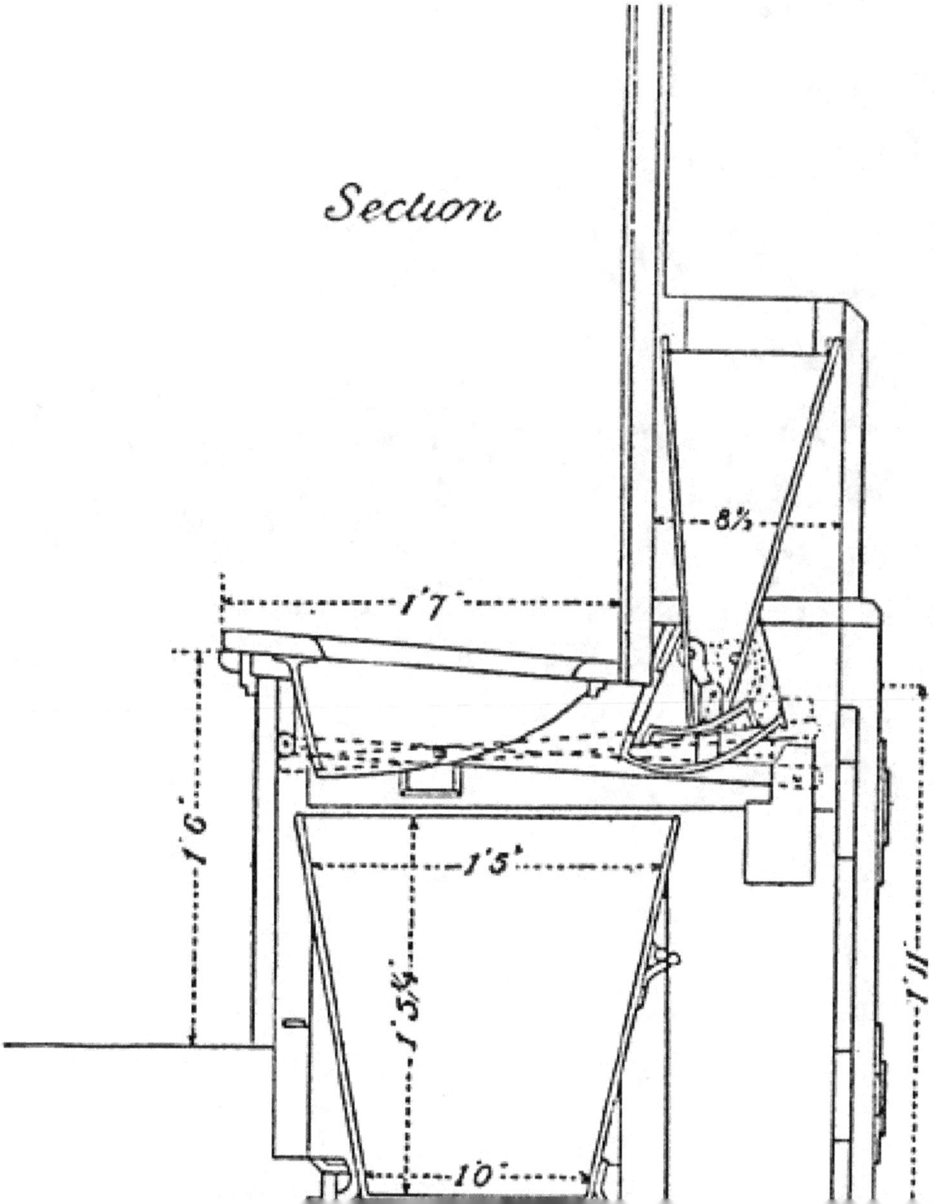

Composting container of "TerraNova" composting toilet, showing open removal chamber (town house at the ecological settlement Hamburg-Allermöhe, Germany)

Chapter 12

Decompiculture

The term **decompiculture** is a neologism coined by forestry professor Timothy Myles of the Urban Entomology Program at the University of Toronto and refers to how decomposing organisms, like termites, could be grown or cultured for a variety of uses.

Myles proposes that people could live in symbiosis with termites by utilizing them in landfills to decompose waste, to improve soils by composting materials, to detoxify hazardous substances, and to produce biomass for animal feed and production of biochemicals. He speculates that decompiculture could eventually become a new biological field that could have significant and important impacts on both humans and termites.

12.1 References

- Decompiculture: Human symbiosis with decomposer organisms

Chapter 13

Decomposition

For other uses, see Decomposition (disambiguation).

Decomposition is the process by which organic substances are broken down into a much simpler form of matter. The process is essential for recycling the finite matter that occupies physical space in the biome. Bodies of living organisms begin to decompose shortly after death. Although no two organisms decompose in the same way, they all undergo the same sequential stages of decomposition. The science which studies decomposition is generally referred to as *taphonomy* from the Greek word τάφος *taphos*, meaning tomb.

One can differentiate **abiotic** from **biotic decomposition** (biodegradation). The former means "degradation of a substance by chemical or physical processes, e.g. hydrolysis."[1] The latter one means "the metabolic breakdown of materials into simpler components by living organisms",[2] typically by microorganisms.

13.1 Animal decomposition

Decomposition begins at the moment of death, caused by two factors: autolysis, the breaking down of tissues by the body's own internal chemicals and enzymes, and putrefaction, the breakdown of tissues by bacteria. These processes release gases that are the chief source of the unmistakably putrid odor of decaying animal tissue.

Prime decomposers are bacteria or fungi, though larger scavengers also play an important role in decomposition if the body is accessible to insects, mites and other animals. The most important arthropods that are involved in the process include carrion beetles, mites,[3] [4] the flesh-flies (Sarcophagidae) and blow-flies (Calliphoridae), such as the greenbottle fly seen in the summer. The most important non-insect animals that are typically involved in the process include mammal and bird scavengers, such as coyotes, dogs, wolves, foxes, rats, crows and vultures. Some of these scavengers also remove and scatter bones, which they ingest at a later time. Aquatic and marine environments have break-down agents that include bacteria, fish, crustaceans, Diptera larvae [5] and other carrion scavengers.

13.1.1 Stages of decomposition

Five general stages are used to describe the process of decomposition in vertebrate animals: fresh, bloat, active and advanced decay, and dry/remains.[6] The general stages of decomposition are coupled with two stages of chemical decomposition: autolysis and putrefaction.[7] These two stages contribute to the chemical process of decomposition, which breaks down the main components of the body.

Ants eating a dead snake

Fresh

The fresh stage begins immediately after the heart stops beating.*[8] From the moment of death, the body begins losing heat to the surrounding environment, resulting in an overall cooling called algor mortis.*[9] Shortly after death, within three to six hours, the muscular tissues become rigid and incapable of relaxing which is known as rigor mortis. Since blood is no longer being pumped through the body it drains to the dependent portions of the body, under gravity, creating an overall bluish-purple discolouration termed livor mortis or, more commonly, lividity.

Once the heart stops, the blood can no longer supply oxygen or remove carbon dioxide from the tissues. The resulting decrease in pH and other chemical changes cause cells to lose their structural integrity, bringing about the release of cellular enzymes capable of initiating the breakdown of surrounding cells and tissues. This process is known as autolysis. Visible changes caused by decomposition are limited during the fresh stage, although autolysis may cause blisters to appear at the surface of the skin.*[10]

The small amount of oxygen remaining in the body is quickly depleted by cellular metabolism and aerobic microbes naturally present in respiratory and gastrointestinal tracts, creating an ideal environment for the proliferation of anaerobic organisms. These multiply, consuming the body's carbohydrates, lipids, and proteins, to produce a variety of substances including propionic acid, lactic acid, methane, hydrogen sulfide and ammonia. The process of microbial proliferation within a body is referred to as putrefaction and leads to the second stage of decomposition, known as bloat.*[8]

Blowflies and flesh flies are the first carrion insects to arrive, and seek a suitable oviposition site.*[6]

Bloat

The bloat stage provides the first clear visual sign that microbial proliferation is underway. In this stage, anaerobic metabolism takes place, leading to the accumulation of gases, such as hydrogen sulfide, carbon dioxide, methane, and nitrogen. The accumulation of gases within the bodily cavity causes the distention of the abdomen and gives a cadaver its overall bloated appearance.[11] The gases produced also cause natural liquids and liquefying tissues to become frothy.[9] As the pressure of the gases within the body increases, fluids are forced to escape from natural orifices, such as the nose, mouth, and anus, and enter the surrounding environment. The buildup of pressure combined with the loss of integrity of the skin may also cause the body to rupture.[11]

Intestinal anaerobic bacteria transform haemoglobin into sulfhemoglobin and other colored pigments. The associated gases which accumulate within the body at this time aid in the transport of sulfhemoglobin throughout the body via the circulatory and lymphatic systems, giving the body an overall marbled appearance.[12]

If insects have access, maggots hatch and begin to feed on the body's tissues.[6] Maggot activity, typically confined to natural orifices and masses under the skin, causes the skin to slip and hair to detach from the skin.[9] Maggot feeding, and the accumulation of gases within the body, eventually leads to post-mortem skin ruptures which will then further allow purging of gases and fluids into the surrounding environment.[8] Ruptures in the skin allow oxygen to re-enter the body and provide more surface area for the development of fly larvae and the activity of aerobic microorganisms.[11] The purging of gases and fluids results in the strong distinctive odours associated with decay.[6]

Active decay

Active decay is characterized by the period of greatest mass loss. This loss occurs as a result of both the voracious feeding of maggots and the purging of decomposition fluids into the surrounding environment.[11] The purged fluids accumulate around the body and create a cadaver decomposition island (CDI).[8] Liquefaction of tissues and disintegration become apparent during this time and strong odours persist.[6] The end of active decay is signaled by the migration of maggots away from the body to pupate.[8]

Advanced decay

Decomposition is largely inhibited during advanced decay due to the loss of readily available cadaveric material.[11] Insect activity is also reduced during this stage.[9] When the carcass is located on soil, the area surrounding it will show evidence of vegetation death.[11] The CDI surrounding the carcass will display an increase in soil carbon and nutrients, such as phosphorus, potassium, calcium, and magnesium;[8] changes in pH; and a significant increase in soil nitrogen.[13]

Dry/remains

During the dry/remains stage, the resurgence of plant growth around the CDI may occur and is a sign that the nutrients present in the surrounding soil have not yet returned to their normal levels.[11] All that remains of the cadaver at this stage is dry skin, cartilage, and bones,[6] which will become dry and bleached if exposed to the elements.[9] If all soft tissue is removed from the cadaver, it is referred to as completely skeletonized, but if only portions of the bones are exposed, it is referred to as partially skeletonised.[14]

13.2   Plant decomposition

See also: Composting, Anaerobic digestion and Fungal extracellular enzyme activity
 Decomposition of plant matter occurs in many stages. It begins with leaching by water; the most easily lost and soluble carbon compounds are liberated in this process. Another early process is physical breakup or fragmentation of the plant material into smaller bits which have greater surface area for microbial colonization and attack. In smaller dead plants, this process is largely carried out by the soil invertebrate fauna, whereas in the larger plants, primarily parasitic life-forms

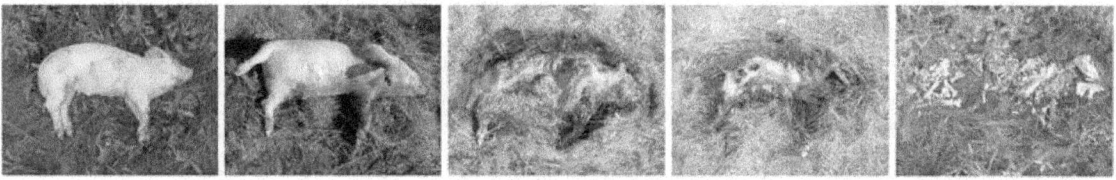

Pig carcass in the different stages of decomposition: Fresh > Bloat > Active decay > Advanced decay > Dry remains

A decaying peach over a period of six days. Each frame is approximately 12 hours apart, as the fruit shrivels and becomes covered with mold.

such as insects and fungi play a major breakdown role and are not assisted by numerous detritivore species. Following this, the plant detritus (consisting of cellulose, hemicellulose, microbial products, and lignin) undergoes chemical alteration by microbes. Different types of compounds decompose at different rates. This is dependent on their chemical structure. For instance, lignin is a component of wood, which is relatively resistant to decomposition and can in fact only be decomposed by certain fungi, such as the black-rot fungi. Said fungi are thought to be seeking the nitrogen content of lignin rather than its carbon content. Lignin is one such remaining product of decomposing plants with a very complex chemical structure causing the rate of microbial breakdown to slow. Warmth determines the speed of plant decay, with the rate of decay increasing as heat increases, i.e. a plant in a warm environment will decay over a shorter period of time. In most grassland ecosystems, natural damage from fire, insects that feed on decaying matter, termites, grazing mammals, and the physical movement of animals through the grass are the primary agents of breakdown and nutrient cycling, while bacteria and fungi play the main roles in further decomposition.

The chemical aspects of plant decomposition always involve the release of carbon dioxide.

13.3   Food decomposition

Main article: Meat spoilage

The decomposition of food, either plant or animal, called *spoilage* in this context, is an important field of study within food science. Food decomposition can be slowed down by conservation. The spoilage of meat occurs, if the meat is untreated, in a matter of hours or days and results in the meat becoming unappetizing, poisonous or infectious. Spoilage is caused by the practically unavoidable infection and subsequent decomposition of meat by bacteria and fungi, which are borne by the animal itself, by the people handling the meat, and by their implements. Meat can be kept edible for a much longer time – though not indefinitely – if proper hygiene is observed during production and processing, and if appropriate food safety, food preservation and food storage procedures are applied.

13.4   Importance to forensics

Further information: Forensic entomological decomposition

Various sciences study the decomposition of bodies under the general rubric of forensics because the usual motive for such studies is to determine the time and cause of death for legal purposes:

- Forensic taphonomy specifically studies the processes of decomposition in order to apply the biological and chemical principles to forensic cases in order to determine post-mortem interval (PMI), post-burial interval as well as to locate clandestine graves.

- Forensic pathology studies the clues to the cause of death found in the corpse as a medical phenomenon.

- Forensic entomology studies the insects and other vermin found in corpses; the sequence in which they appear, the kinds of insects, and where they are found in their life cycle are clues that can shed light on the time of death, the length of a corpse's exposure, and whether the corpse was moved.[15][16]

- Forensic anthropology is the branch of physical anthropology that studies skeletons and human remains, usually to seek clues as to the identity, race, and sex of their former owner.[17][18]

The University of Tennessee Anthropological Research Facility (better known as the Body Farm) in Knoxville, Tennessee has a number of bodies laid out in various situations in a fenced-in plot near the medical center. Scientists at the Body Farm study how the human body decays in various circumstances to gain a better understanding of decomposition.

13.5   Factors affecting decomposition

Further information: Environmental effects on forensic entomology

13.5.1 Exposure to the elements

A dead body that has been exposed to the open elements, such as water and air, will decompose more quickly and attract much more insect activity than a body that is buried or confined in special protective gear or artifacts. This is due, in part, to the limited number of insects that can penetrate a coffin and the lower temperatures under soil.

The rate and manner of decomposition in an animal body is strongly affected by several factors. In roughly descending degrees of importance, they are:

- Temperature;

- The availability of oxygen;

- Prior embalming;

- Cause of death;

- Burial, depth of burial, and soil type;

- Access by scavengers;

- Trauma, including wounds and crushing blows;

- Humidity, or wetness;

- Rainfall;

- Body size and weight;

- Clothing;

- The surface on which the body rests;

- Foods/objects inside the specimen's digestive tract (bacon compared to lettuce).

The speed at which decomposition occurs varies greatly. Factors such as temperature, humidity, and the season of death all determine how fast a fresh body will skeletonize or mummify. A basic guide for the effect of environment on decomposition is given as Casper's Law (or Ratio): if all other factors are equal, then, when there is free access of air a body decomposes twice as fast than if immersed in water and eight times faster than if buried in earth. Ultimately, the rate of bacterial decomposition acting on the tissue will be depend upon the temperature of the surroundings. Colder temperatures decrease the rate of decomposition while warmer temperatures increase it.

The most important variable is a body's accessibility to insects, particularly flies. On the surface in tropical areas, invertebrates alone can easily reduce a fully fleshed corpse to clean bones in under two weeks. The skeleton itself is not permanent; acids in soils can reduce it to unrecognizable components. This is one reason given for the lack of human remains found in the wreckage of the *Titanic*, even in parts of the ship considered inaccessible to scavengers. Freshly skeletonized bone is often called "green" bone and has a characteristic greasy feel. Under certain conditions (normally cool, damp soil), bodies may undergo saponification and develop a waxy substance called adipocere, caused by the action of soil chemicals on the body's proteins and fats. The formation of adipocere slows decomposition by inhibiting the bacteria that cause putrefaction.

In extremely dry or cold conditions, the normal process of decomposition is halted – by either lack of moisture or temperature controls on bacterial and enzymatic action – causing the body to be preserved as a mummy. Frozen mummies commonly restart the decomposition process when thawed (see Ötzi the Iceman), whilst heat-desiccated mummies remain so unless exposed to moisture.

The bodies of newborns who never ingested food are an important exception to the normal process of decomposition. They lack the internal microbial flora that produce much of decomposition and quite commonly mummify if kept in even moderately dry conditions.

13.5.2 Artificial preservation

Embalming is the practice of delaying decomposition of human and animal remains. Embalming slows decomposition somewhat, but does not forestall it indefinitely. Embalmers typically pay great attention to parts of the body seen by mourners, such as the face and hands. The chemicals used in embalming repel most insects, and slow down bacterial putrefaction by either killing existing bacteria in or on the body themselves or by "fixing" cellular proteins, which means that they cannot act as a nutrient source for subsequent bacterial infections. In sufficiently dry environments, an embalmed body may end up mummified and it is not uncommon for bodies to remain preserved to a viewable extent after decades. Notable viewable embalmed bodies include those of:

- Eva Perón of Argentina, whose body was injected with paraffin was kept perfectly preserved for many years, and still is as far as is known (her body is no longer on public display).

- Vladimir Lenin of the Soviet Union, whose body was kept submerged in a special tank of fluid for decades and is on public display in Lenin's Mausoleum.

  - Other Communist leaders with pronounced cults of personality such as Mao Zedong, Kim Il-sung, Ho Chi Minh, Kim Jong-il and most recently Hugo Chávez have also had their cadavers preserved in the fashion of Lenin's preservation and are now displayed in their respective mausoleums.

- Pope John XXIII, whose preserved body can be viewed in St. Peter's Basilica.

- Padre Pio, whose body was injected with formalin prior to burial in a dry vault from which he was later removed and placed on public display at the San Giovanni Rotondo.

13.5.3 Environmental preservation

A body buried in a sufficiently dry environment may be well preserved for decades. This was observed in the case for murdered civil rights activist Medgar Evers, who was found to be almost perfectly preserved over 30 years after his death, permitting an accurate autopsy when the case of his murder was re-opened in the 1990s.[19]

Bodies submerged in a peat bog may become naturally "embalmed", arresting decomposition and resulting in a preserved specimen known as a bog body. The time for an embalmed body to be reduced to a skeleton varies greatly. Even when a body is decomposed, embalming treatment can still be achieved (the arterial system decays more slowly) but would not restore a natural appearance without extensive reconstruction and cosmetic work, and is largely used to control the foul odors due to decomposition.

An animal can be preserved almost perfectly, for millions of years in a resin such as amber.

There are some examples where bodies have been inexplicably preserved (with no human intervention) for decades or centuries and appear almost the same as when they died. In some religious groups, this is known as incorruptibility. It is not known whether or for how long a body can stay free of decay without artificial preservation.[20]

13.6 See also

- Cadaverine

- Decompiculture

- Microbiology of decomposition

- Putrescine

- Staling

- Humus

- Leachate

- Peat (turf)

13.7 References

[1] Water Quality Vocabulary. ISO 6107-6:1994.

[2] Water Words Dictionary (WWD)

[3] González Medina A, González Herrera L, Perotti MA, Jiménez Ríos G (2013). "Occurrence of *Poecilochirus austroasiaticus* (Acari: Parasitidae) in forensic autopsies and its application on postmortem interval estimation". *Exp.Appl.Acarol.* **59** (3): 297–305. doi:10.1007/s10493-012-9606-1.

[4] Braig, Henk R.; Perotti, M. Alejandra (2009). "Carcases and mites". *Experimental and Applied Acarology* **49** (1–2): 45–84. doi:10.1007/s10493-009-9287-6.

[5] González Medina A, Soriano Hernando Ó, Jiménez Ríos G (2015). "The Use of the Developmental Rate of the Aquatic Midge *Chironomus riparius* (Diptera, Chironomidae) in the Assessment of the Postsubmersion Interval". *J.Forensic.Sci* **60** (3): 822–826. doi:10.1111/1556-4029.12707.

[6] Payne, J.A. (1965). "A summer carrion study of the baby pig sus scrofa Linnaeus". *Ecology* **46** (5): 592–602. doi:10.2307/1934999.

[7] Forbes, S.L. (2008). "Decomposition Chemistry in a Burial Environment". In M. Tibbett, D.O. Carter. *Soil Analysis in Forensic Taphonomy*. CRC Press. pp. 203–223. ISBN 1-4200-6991-8.

[8] Carter D.O., Yellowlees, D., Tibbett M. (2007). "Cadaver decomposition in terrestrial ecosystems". *Naturwissenschaften* **94** (1): 12–24. doi:10.1007/s00114-006-0159-1. PMID 17091303.

[9] Janaway R.C., Percival S.L., Wilson A.S. (2009). "Decomposition of Human Remains". In Percival, S.L. *Microbiology and Aging*. Springer Science + Business. pp. 13–334. ISBN 1-58829-640-7.

[10] Knight, Bernard (1991). *Forensic pathology*. Oxford University Press. ISBN 0-19-520903-6.

[11] Carter D.O., Tibbett M. (2008). "Cadaver Decomposition and Soil: Processes". In M. Tibbett, D.O. Carter. *Soil Analysis in Forensic Taphonomy*. CRC Press. pp. 29–51. ISBN 1-4200-6991-8.

[12] Pinheiro, J. (2006). "Decay Process of a Cadaver". In A. Schmidt, E. Cumha, J. Pinheiro. *Forensic Anthropology and Medicine*. Humana Press. pp. 85–116. ISBN 1-58829-824-8.

[13] Vass A.A., Bass W.M., Wolt J.D., Foss J.E., Ammons J.T. (1992). "Time since death determinations of human cadavers using soil solution". *Journal of Forensic Sciences* **37** (5): 1236–1253. PMID 1402750.

[14] Dent B.B., Forbes S.L., Stuart B.H. "Review of human decomposition processes in soil". *Environmental Geology* **45**: 576–585. doi:10.1007/s00254-003-0913-z.

[15] Smith, KGV. (1987). *A Manual of Forensic Entomology*. Cornell Univ. Pr. p. 464. ISBN 0-8014-1927-1.

[16] Kulshrestha P, Satpathy DK. (2001). "Use of beetles in forensic entomology". *Forensic Sci. Int.* **120** (1–2): 15–17. doi:10.1016/S0379-0738(01)00410-8. PMID 11457603.

[17] Schmitt, A.; Cunha, E.; Pinheiro, J. (2006). *Forensic Anthropology and Medicine: Complementary Sciences From Recovery to Cause of Death*. Humana Press. p. 464. ISBN 1-58829-824-8.

[18] Haglund, WD.; Sorg, MH. (1996). *Forensic Taphonomy: The Postmortem Fate of Human Remains*. CRC Press. p. 636. ISBN 0-8493-9434-1.

[19] Quigley, C. (1998). *Modern Mummies: The Preservation of the Human Body in the Twentieth Century*. McFarland. pp. 213–214. ISBN 0-7864-0492-2.

[20] Clark, Josh. "How can a corpse be incorruptible?". HowStuffWorks.

13.8 External links

- Media related to Decomposition at Wikimedia Commons

- 1Lecture.com – Food decomposition (a Flash animation)

Chapter 14

Dillo Dirt

Dillo Dirt is a compost made by the City of Austin, Texas since 1989. It was the first program of its kind in the state and one of the oldest in the nation.[*][1] Dillo Dirt is named after the Nine-banded Armadillo *(Dasypus novemcinctus)*, which is a mammal native to Texas. It is also a trademarked product of the City of Austin Water Department.

The unique difference between Dillo Dirt and normal compost is that it contains treated municipal sewage sludge along with yard trimmings collected curbside by the City of Austin Resource Recovery Department. These are combined and composted to create Dillo Dirt. Despite this fact, Dillo Dirt meets all Texas and U.S. Environmental Protection Agency requirements for "unrestricted" use, which even includes vegetable gardens.

The heat generated in composting (130 to 170 °F (50 to 80 °C)) is sufficient to virtually eliminate human and plant pathogens. After active composting for over a month, the compost is "cured" for several months, then screened to produce the finished product.

According to the City of Austin, Dillo Dirt contains levels of heavy metals including Arsenic, Cadmium, Copper, Lead, Mercury, Molybdenum, Nickel, Selenium, and Zinc. In a separate toxicological analysis of Dillo Dirt, levels of the following pollutants were found: Beta-BHC, DDE, Dieldrin, Endrin aldehyde, Benzo(b)fluoranthene, Dibenz(a,h)anthracene, Benzo(a)anthracene, Indeno(1,2,3-cd)pyrene, and Bis(2-ethylhexyl)phthalate. Very few tests have been carried out on Dillo Dirt, so average pollutant, radioactivity, or carcinogen levels are generally unknown.[*][2]

14.1   Controversy

Some opponents of the use and sale of Dillo Dirt claim that it contains above-normal amounts of heavy metals and fluoride which will inevitably find their way back into the human food supply.[*][3] The city, however, states that the metal levels are well below the federal allowable levels.[*][4]

Prior to the Austin City Limits Music Festival, held at Zilker Park in Austin on October 4, 2009, the park's soil was resurfaced and amended with Dillo Dirt. During the festival, heavy rains created a large amount of sludgy surface mud, which some concertgoers claim caused them health issues, such as skin rashes,[*][5] but in the end no conclusive evidence that the Dillo Dirt caused any problems was determined.[*][6] The local Austin American Statesman newspaper ran an article advising anyone affected to call the health department.

14.2   See also

- Dillo Dirt FAQ at the City of Austin website

- Uses and Application Rates of Dillo Dirt

14.3   References

[1]   "Introduction to Dillo Dirt" .

[2]   "Austin's Dirty Secret: Dillo Dirt" .

[3]   "AUSTIN'S DISTURBING DILLO DIRT: ANOTHER WAY TO SELL TOXIC WASTE" .

[4]   "Dillo Dirt FAQ" .

[5]   "Austin's Dirty Secret: Dillo Dirt" .

[6]   "Dillo Dirt rash? Call health department at 972-5555" .

Chapter 15

Ecuador composting method

A mud layer being placed on a layer of organic biomass, Monte de los Olivos, Ucayali, Peru

The Ecuador composting method is a common composting practice in the lowlands of Ecuador and Peru. The compost pile is embedded on the tree trunk or banana stalks, with a pale erected in the middle. Organic matter is placed in layers on the trunks or stalks, each layer being covered by mud, or inlaid via different types of organic matter. When the pile is about 1.2 m high, it is watered and covered by big leaves. After some time, when the compost pile settles down, the central pale is removed for aeration. This composting method is typically done in a small-scale, by indigenous villagers.

Compost pile covered by leaves of banana (Musa), *Santa Tereza, Ucayali, Peru*

Chapter 16

Eisenia fetida

Eisenia fetida (older spelling: foetida), known under various common names such as *redworm, brandling worm, panfish worm, trout worm, tiger worm, red wiggler worm, red californian earth worm*, etc., is a species of earthworm adapted to decaying organic material. These worms thrive in rotting vegetation, compost, and manure. They are epigean, rarely found in soil. In this trait they resemble *Lumbricus rubellus*.

They have groups of bristles (called setae) on each segment that move in and out to grip nearby surfaces as the worms stretch and contract their muscles to push themselves forward or backward.

Eisenia fetida worms are used for vermicomposting. They are native to Europe, but have been introduced (both intentionally and unintentionally) to every other continent except Antarctica.

16.1   Odor

When roughly handled, an *Eisenia fetida* exudes a pungent liquid, thus the specific name *foetida* meaning *foul-smelling*. This is presumably an antipredator adaptation.

16.2   Related species

Eisenia fetida is closely related to *Eisenia andrei*, also referred to as *E. foetida andrei*. The only simple way of distinguishing the two species is that *E. foetida* is sometimes lighter in colour. Molecular analyses have confirmed their identity as separate species and breeding experiments have shown that they do not produce hybrids.

16.3   Reproduction

As with other earthworm species, *Eisenia fetida* is hermaphroditic. However, two worms are still required for reproduction. The two worms join clitella, the large lighter-colored bands which contain the worms' reproductive organs, and which are only prominent during the reproduction process. The two worms exchange sperm. Both worms then secrete cocoons which contain several eggs each. These cocoons are lemon-shaped and are pale yellow at first, becoming more brownish as the worms inside become mature. These cocoons are clearly visible to the naked eye.

16.4   References

[1] "*Eisenia foetida*". Fauna Europaea. 2004.

Close-up of Eisenia fetida *with visible bristles*

Chapter 17

Fairfield Materials Management Ltd

Fairfield Materials Management Ltd is a Manchester based social enterprise that operates a community waste management project at Manchester's New Smithfield Market focused on minimising waste, and bringing social and environmental benefits to Greater Manchester.[1]

By utilising 'in-vessel composting', Fairfield Materials Management established the UK's first sustainable biodegradable waste management system to operate on a wholesale market,[2] diverted 16,500 tonnes of organic market waste material away from landfill between 2003 and 2008.[3]

Fairfield has developed a composting production model that processes fruit, vegetable, plant and woody waste into peat-free, British Standards Institution PAS 100[4] accredited compost.[5] It operates on a site fully licensed by the Environment Agency.[6]

In 2003 Fairfield was recognised as the most innovative social enterprise within its sector by The Composting Association.[7]

17.1   History

Fairfield Materials Management was founded in 2003 by a small group of ecological activists, horticulturists and social entrepreneurs.[8]

Fairfield Materials Management was initiated by Val Rawlinson (Director) after she became involved in an East Manchester anti-incineration campaign in 1996. The incinerator was not built, but the campaigners realised that they had to show Manchester that environmental and social alternatives were available and working successfully throughout the United Kingdom and so Fairfield Composting (later to be called Debdale Eco Centre) was established.[9]

Val recognised the need for a commercial sized composting system for New Smithfield Market and choose a Vertical Composting Unit (VCU) system as the preferred technology; Emma Smith was recruited as a waste auditor in November 2001 to undertake a waste composition analysis of Smithfield Market's waste. Chris Walsh joined the team and between 2001 and 2003 the group secured several hundred thousand pounds through grants and loans,[10] designed the site, obtaining planning permission and a waste management licence.[11]

In 2003 Fairfield began operating and initially diverted waste from four market traders, taking in green waste from Manchester City Council and producing compost. In 2004 a further two composting units were added and in 2005 a final three units were added to the system. This enabled Fairfield to process all of the market's fruit and vegetable waste through the in-situ technology.[12]

In 2009 the Fairfield Group in partnership national renewable energy business Bio Group Ltd [13] secured planning permission in Bredbury, Stockport for their second waste management site.[14]

17.2   Aims

17.3   References

[1] http://www.fairfieldcompost.co.uk/about/history.html

[2] http://www.fairfieldcompost.co.uk/about.html

[3] http://www.fairfieldcompost.co.uk/what_we_do.html

[4] http://www.wrap.org.uk/composting/compost_ specifications/bsi_pas_100/bsi_pas_100_1.

[5] http://www.letsrecycle.com/materials/composting/news.jsp?story=5785

[6] http://www.fairfieldcompost.co.uk/about.html

[7] http://www.compostingconference.com/Community_Init_Award.html

[8] http://www.fairfieldcompost.co.uk/about.html

[9] http://www.debdale-eco-centre.com

[10] http://www.adventurecapitalfund.org.uk/content/view/28/59/

[11] http://www.communitycompost.org/hotrotters/fmm.htm

[12] http://www.fairfieldcompost.co.uk/about/history.html

[13] http://www.crainsmanchesterbusiness.co.uk/apps/pbcs.dll/article?AID=/20090106/FREE/301069997/1046/energy

[14] http://www.englandsnorthwest.com/invest/news/archive/8162-food-recycle-plant-approved.html

17.4   External links

- Fairfield Materials Management
- Community Compost Network
- Fairfield AD Ltd

Chapter 18

Grasscycling

Grasscycling refers to an aerobic method *[1] of handling grass clippings by leaving them on the lawn when mowing.

Electric lawn mower in grass-cycling mode

The term is a portmanteau combining "grass" and "recycling", and had come into use by at least 1990*[2] as part of the push to reduce the huge quantities of clippings going into landfills, up to half of some cities' summertime waste flow,[3] as 1,000 square feet (93 m^2) of lawn can produce 200 to 500 pounds (90 to 225 kg) of clippings a year.[4]

Because grass consists largely of water (80% or more[5]), contains little lignin,[5] and has high nitrogen content, grass clippings easily break down during an aerobic process [1] (comparable to composting) and returns the decomposed clippings to the soil within one to two weeks,[4] acting primarily as a fertilizer supplement and, to a much smaller degree, a mulch. Grasscycling can provide 15 to 20% or more of a lawn's yearly nitrogen requirements.[6][7] Proponents also note that grasscycling reduces the use of plastic bags for collecting yard waste and reduces trips to the curb or landfill to haul waste.[8]

Optimal grasscycle techniques include:[4][6][9]

- Cutting no more than 1/3 the length of the grass

- Cutting when the grass is dry to the touch

- Cutting when the height is between 3 and 4 inches (7 to 10 cm)

- Ensuring that the mower blade is sharp

Although a mulching mower can make grass clippings smaller, one is not necessary for grasscycling.

18.1   See also

- Compost

- Recycling

- List of organic gardening and farming topics

18.2   References

[1] Scoville, Heather. "Aerobic vs. Anaerobic Processes" . *About Education*. About.com. Retrieved April 17, 2015.

[2] "Grasscycling definition/etymology" . Retrieved 2007-05-25.

[3] "Denver Recycle Grasscycle" . Retrieved 2007-05-25.

[4] "Rivanna Solid Waste Authority Grasscycling Info" . Archived from the original on 2007-09-30. Retrieved 2007-05-25.

[5] "Grasscycling FAQ" . Retrieved 2007-05-25.

[6] "Grasscycle!". Retrieved 2007-05-25.

[7] "California Integrated Waste Management Board - What is grasscycling?". Retrieved 2007-05-25.

[8] "King County Experience: Grasscycling" . Archived from the original on 1999-11-05. Retrieved 2007-05-25.

[9] "Bay Delta Grass Recycling Campaign" . Retrieved 2007-05-25.

18.3   External links

- What is grasscycling?

Chapter 19

Hermetia illucens

Hermetia illucens, the **black soldier fly**, is a common and widespread fly of the family Stratiomyidae, whose larvae are common detritivores in compost heaps. Larvae are also sometimes found in association with carrion, and have significant potential for use in forensic entomology.[*][1]

Black soldier fly larvae (BSFL), also known as "phoenix worms", may be used in manure management, for house fly control and for the bioconversion of organic waste material. Mature larvae and prepupae raised in manure management and waste bioconversion operations may also be used to supplement animal feeds.[*][2]

Larvae are sold as feeders for owners of herptiles and tropical fish, or as composting grubs. They store high levels of calcium for future pupation which is beneficial to herptiles.[*][3]

The adult soldier fly has no functioning mouthparts; it spends its time searching for mates and reproducing.

19.1   Life cycle

Black soldier fly eggs take approximately four days to hatch and are typically deposited in crevices or on surfaces above or adjacent to decaying matter such as manure or compost.[*][4] The larvae range in size from $\frac{1}{8}$–$\frac{3}{4}$ inch (3–19 mm). Although they can be stored at room temperature for several weeks, their longest shelf life is achieved at 50–60 °F (10–16 °C).

The adult fly, which measures about 16 mm (5/8 inch),[*][5] has a life span of 5 to 8 days. It is a mimic, very close in size, color, and appearance to the organ pipe mud dauber wasp and its relatives. The mimicry of this particular kind of wasp

77

is especially enhanced in that the fly's antennae are elongated and wasp-like, the fly's hind tarsi are pale, as are the wasp's, and the fly has two small transparent "windows" in the basal abdominal segments that make the fly appear to have a narrow "wasp waist". The adult soldier fly has no functioning mouthparts; it spends its time searching for mates and reproducing.

19.2   Uses in composting or as food for animals

Black soldier fly larvae (BSFL) are used to compost and sanitize wastes, and/or convert the wastes into animal feed. The harvested pupae and prepupae are eaten by poultry, fish, pigs, turtles; even dogs. The wastes include fresh manure, food wastes of both animal and vegetable origin.

19.2.1   Grub composting bins use self-harvesting

When the larvae have completed their larval development through six instars, they enter a stage called the "**prepupa**" wherein they cease to eat, they empty their guts, their mouth parts change to an appendage that aids climbing, and they seek a dry, sheltered area to pupate.[6] This prepupal migration instinct is used by grub composting bins to self-harvest the mature larvae. These containers have ramps or holes on the sides to allow the prepupae to climb out of the composter and drop into a collection area.

19.2.2   Benefits

Larvae are beneficial in the following ways:

- They prevent houseflies and blowflies from laying eggs in the material inhabited by black soldier fly larvae.[7]

- They are usually not a pest.

   - They are not attracted to human habitation or foods.[7] As a detritivore and coprovore, the egg-bearing females are attracted to rotting food or manure.

   - Black soldier flies do not fly around as much as houseflies. They are very easy to catch and relocate when they get inside a house, as they do not avoid being picked up, they are sanitary, and they do not bite or sting. Their only defense seems to be hiding. When using a wet grub bin that will collect or kill all the pupae, the black soldier fly population is easy to reduce by killing the pupae/prepupae in the collection container, before they become flies. They may be killed by freezing, drying, manually feeding to domestic animals, putting the collection container in a chicken coop for automatic feeding, or feeding to wild birds with a mouse/pest-proof feeder.[8]

- Significant reductions of *E. coli* 0157:H7 and *Salmonella enterica* were measured in hen manure.[6]

- They quickly reclaim would-be pollutants: Nine stinky organic chemicals were greatly reduced or eliminated from manure in 24 hours.[6]

- They quickly reduce the volume and weight of would-be waste: The larval colony breaks apart its food, churns it, and creates heat, increasing compost evaporation. Significant amounts are also converted to carbon dioxide respired by the grubs and symbiotic/mutualistic microorganisms.

19.2.3   Establishing and building larval colonies

The main difficulty is obtaining black soldier fly larvae or eggs to start or replenish the colony. This is usually done by enticing the soldier flies to lay eggs in small holes over the grub bin. Adult flies lay clusters of eggs in the edges of corrugated cardboard or corrugated plastic. In some regions, it is possible to start or maintain adequate larvae colonies from native soldier flies; however, pest species such as houseflies and blowflies are also drawn to many of the foods used to attract soldier flies (such as fermented chicken feed).

In tropical or subtropical climates, they might breed year-round, but in other climates, a greenhouse may be needed to obtain eggs in the cooler periods. The grubs are quite hardy and can handle more acidic conditions and higher temperatures than redworms. Larvae can survive cold winters, particularly with large numbers of grubs, insulation, or compost heat (generated by the microorganisms in the grub bin or compost pile). Heat stimulates the grubs to crawl off, pupate, and hatch, and a great deal of light and heat seem to be required for breeding. Many small-scale grub farmers build their larval colonies from eggs deposited by wild soldier flies.

Requirements for captive-breeding chambers

Captive breeding can also keep pest flies away if done carefully.

Space and shape   Newly emerged soldier flies mate in flight.[*][9]

"Tingle et al (1975)···reported that mating and oviposition were observed "often" in a 3 × 6.1 × 1.8 m cage held outdoors. In addition, mating was observed in a 0.76 × 1.14 × 1.37 m cage held outdoors, but not when held in the greenhouse." [*][9]

"No mating or egg collections occurred in two small cages (53×91×53 cm) and (38×46×38 cm)" [*][9]

More recently, captive breeding has been observed in a cylindrical chamber measuring 46 cm (18 in) diameter by 56 cm (22 in) tall for 99 litres (3.5 cubic feet). "Mating under artificial light did not occur but did succeed with natural sunlight." [*][10]

German scientists have successfully bred soldier flies in a space as small as 10 liters.[*][11][*][12]

Heat   "Adults typically mated and oviposited at temperatures of 24 °C (75 °F) up to 40 °C (104 °F) or more. Booth and Sheppard (1984) reported that 99.6% of oviposition in the field occurred at 27.5–37.5 °C (81.5–99.5 °F)" [*][9]

Light   "Minimum light intensity for mating is 63 μ mol m2s -1 with most mating occurring at over 200 μ mol m2s -1 (J.K.T. and D.C.S., unpublished data)." [*][9]

Humidity   "Relative humidities of 30–90% supported mating and oviposition" [*][9]

Human food

Black soldier fly larvae are edible to humans. The larvae are highly efficient in converting proteins, containing up to 42% of protein, and a lot of calcium and amino acids. In 432 hours, 1 gram of black soldier fly eggs turns into 2.4 kilograms of protein. They thus can be a source of protein for human consumption.

In 2013, Austrian designer Katharina Unger invented a table-top insect breeding farm called "Farm 432" in which people can produce edible fly larvae at home. It is a multichambered plastic machine that looks like a kitchen appliance. According to Unger: "Farm 432 enables people to turn against the dysfunctional system of current meat production by growing their own protein source." About 500 g of larvae or two meals can be produced in a week by the machine.

The taste of the larvae is said to be very distinctive. Unger: "When you cook them, they smell a bit like cooked potatoes. The consistency is a bit harder on the outside and like soft meat on the inside. The taste is nutty and a bit meaty." [*][13]

19.2.4   Black soldier fly larvae and redworms

Worm farmers often get larvae in their worm bins. Larvae are best at quickly converting "high-nutrient" waste into animal feed.[*][14] Redworms are better at converting high-cellulose materials (paper, cardboard, leaves, plant materials except wood) into an excellent soil amendment.

Redworms thrive on the residue produced by the fly larvae, but larvae leachate ("tea") contains enzymes and tends to be too acidic for worms. The activity of larvae can keep temperatures around a 100°F, while redworms require cooler temperatures. Most attempts to raise large numbers of larvae with redworms in the same container, at the same time, are unsuccessful. Worms have been able to survive in/under grub bins when the bottom is the ground. Redworms can live in grub bins when a large number of larvae are not present. Worms can be added if the larval population gets low (in the cold season) and worms can be raised in grub bins while awaiting eggs from wild black soldier flies.

As a feeder species, BSFL are not known to be intermediate hosts of parasitic worms that infect poultry, while redworms are host to many.*[15]

19.3   Names and trademarks

A generic term for grubs of *Hermetia illucens* is **black soldier fly larvae**, abbreviated as "BSFL". Black soldier fly larvae were developed as a feeder insect for exotic pets by D. Craig Sheppard. Dr. Sheppard named the larvae **Phoenix Worms** and began marketing them as pet food. In 2006, **Phoenix Worms** became the first feeder insect to be granted a U.S. registered trademark. Other companies also market black soldier fly larvae and use their own brand names such as "**Soldier Grubs**," "**Reptiworms**" and "**Calciworms**." In Australia Black soldier Fly Larvae are marketed as live feeder insects under the trade marked brand name "Beardie Grubs" at selected pet stores.

19.4   References

[1] Lord, W. D., Goff, M. L., Adkins, T. R., and Haskell, N. H. (1994). "The black soldier fly *Hermetia illucens* (Diptera: Stratiomyidae) as a potential measure of human postmortem interval: observations and case histories". *Journal of Forensic Sciences* **39** (1): 215–222. doi:10.1520/JFS13587J. PMID 8113702.

[2] Sheppard, D. C. (1992). "Large-scale Feed Production from Animal Manures with a Non-Pest Native Fly". *Food Insects Newsletter* **5** (2).

[3] "The Incredible Edible Worm", by Audrey Pavia, Reptiles Magazine, July, 2007

[4] *Hermetia illucens* (University of Florida)

[5] Savonen, Carol. "Big maggots in your compost? They're soldier fly larvae". *OSU Extension Service - Gardening.* Oregon State University.

[6] "Research Summary: Black Soldier Fly Prepupae - A Compelling Alternative to Fish Meal and Fish Oil". February 14, 2011.

[7] "Black Soldier Fly: Compiled Research On Best Cultivation Practices". Research Resources. 9 July 2008.

[8] Feeding Grubs to Birds EXPERIMENT

[9] "BSD Prepares to Test Soldier Fly Mating Facility (follow up comment)".

[10] Black Soldier Fly Indoor Breeding Inclosure

[11] Breeding BSF in captivity / Re: not easy

[12] (translation of) Zucht der schwarzen Soldatenfliege (Hermetia illucens)

[13] Katharina Unger, "Farm 432: Insect Breeding" in *Dezeen Magazine*, 25.3.2013, at http://www.dezeen.com/2013/07/25/farm-432-insect-breeding-kitchen-appliance-by-katharina-unger/

[14] Modern animal feed eco-friendly solutions

[15] "TABLE 05: Common Helminths of Poultry". *The Merck Veterinary Manual / Poultry / Helminthiasis.* Retrieved April 20, 2008.

19.5   External links

- Black Soldier Fly Blog
- Black Soldier Fly Farming
- The GrubCycle Solution
- Forging a New Food Chain
- Report for Mike Williams (2005)
- Bioconversion of Food Waste : Black Soldier fly
- 'Grubby' Research Promises Environmental, Economic Benefits
- Black soldier fly on the UF / IFAS *Featured Creatures* website

Female black soldier fly depositing eggs in tree bark.

Black soldier fly depositing eggs in cardboard

Black soldier fly inflating its wings during the first 15 minutes after emergence from pupation.

Chapter 20

Hotbed

In **biology**, a **hotbed** is a pile of decaying organic matter warmer than its surroundings due to the heat given off by the metabolism of the microorganisms in the decomposing pile.

A hotbed covered with a small glass cover (also called a **hotbox**) is used as a small version of a hothouse (heated greenhouse). The bed is often made of manure from animals such as horses that pass much undigested plant cellulose in their droppings. Thus hotboxes are to cold frames what hothouses are to greenhouses.

Some egg-laying animals make or use hotbeds to incubate their eggs: for example the brush turkey.

Chapter 21

Humic acid

Humic acid is a principal component of humic substances, which are the major organic constituents of soil (humus), peat, coal, many upland streams, dystrophic lakes, and ocean water.[*][1] It is produced by biodegradation of dead organic matter. It is not a single acid; rather, it is a complex mixture of many different acids containing carboxyl and phenolate groups so that the mixture behaves functionally as a dibasic acid or, occasionally, as a tribasic acid. Humic acids can form complexes with ions that are commonly found in the environment creating humic colloids. Humic and **fulvic acids** (fulvic acids are humic acids of lower molecular weight and higher oxygen content than other humic acids) are commonly used as a soil supplement in agriculture, and less commonly as a human nutritional supplement. As a nutrition supplement, fulvic acid can be found in a liquid form as a component of mineral colloids. Fulvic acids are poly-electrolytes and are unique colloids that diffuse easily through membranes whereas all other colloids do not.[*][2]

21.1 Formation and description

The formation of humic substances is one of the least understood aspects of humus chemistry and one of the most intriguing. There are three main theories to explain it: the lignin theory of Waksman (1932), the polyphenol theory, and the sugar-amine condensation theory of Maillard (1911).[*][3][*][4]

Humic substances are formed by the microbial degradation of dead plant matter, such as lignin. They are very resistant to further biodegradation. The precise properties and structure of a given sample depend on the water or soil source and the specific conditions of extraction. Nevertheless, the average properties of humic substances from different sources are remarkably similar.

Humic substances in soils and sediments can be divided into three main fractions: humic acids, fulvic acids, and humin. The humic and fulvic acids are extracted as a colloidal sol from soil and other solid phase sources into a strongly basic aqueous solution of sodium hydroxide or potassium hydroxide. Humic acids are precipitated from this solution by adjusting the pH to 1 with hydrochloric acid, leaving the fulvic acids in solution. This is the operational distinction between humic and fulvic acids. Humin is insoluble in dilute alkali. The alcohol-soluble portion of the humic fraction is, in general, named *ulmic acid*. So-called "gray humic acids" (GHA) are soluble in low-ionic-strength alkaline media; "brown humic acids" (BHA) are soluble in alkaline conditions independent of ionic strength; and fulvic acids (FA) are soluble independent of pH and ionic strength.[*][5]

Liquid chromatography and liquid-liquid extraction can be used to separate the components that make up a humic substance. Substances identified include mono-, di-, and tri-hydroxy acids, fatty acids, dicarboxylic acids, linear alcohols, phenolic acids, and terpenoids.[*][6]

Example of a typical humic acid, having a variety of components including quinone, phenol, catechol and sugar moieties[1]

21.2   Chemical characteristics of humic substances

A typical humic substance is a mixture of many molecules, some of which are based on a motif of aromatic nuclei with phenolic and carboxylic substituents, linked together; the illustration shows a typical structure. The functional groups that contribute most to surface charge and reactivity of humic substances are phenolic and carboxylic groups.*[1] Humic acids behave as mixtures of dibasic acids, with a pK_1 value around 4 for protonation of carboxyl groups and around 8 for protonation of phenolate groups. There is considerable overall similarity among individual humic acids.*[7] For this reason, measured pK values for a given sample are average values relating to the constituent species. The other important characteristic is charge density. The molecules may form a supramolecular structure held together by non-covalent forces, such as Van der Waals force, π-π, and CH-π bonds.*[8]

The presence of carboxylate and phenolate groups gives the humic acids the ability to form complexes with ions such as Mg^*2+, Ca^*2+, Fe^*2+ and Fe^*3+. Many humic acids have two or more of these groups arranged so as to enable the formation of chelate complexes.*[9] The formation of (chelate) complexes is an important aspect of the biological role of humic acids in regulating bioavailability of metal ions.*[7]

21.3   Determination of humic acids in water samples

The presence of humic acid in water intended for potable or industrial use can have a significant impact on the treatability of that water and the success of chemical disinfection processes. Accurate methods of establishing humic acid concentrations are therefore essential in maintaining water supplies, especially from upland peaty catchments in temperate climates.

As a lot of different bio-organic molecules in very diverse physical associations are mixed together in natural environments, it is cumbersome to measure their exact concentrations in the humic superstructure. For this reason, concentrations of humic acid are traditionally estimated out of concentrations of organic matter (typically from concentrations of total organic carbon (TOC) or dissolved organic carbon (DOC).

Extraction procedures are bound to alter some of the chemical linkages present in the soil humic substances (mainly ester bonds in biopolyesters such as cutins and suberins). The humic extracts are composed of large numbers of different bio-organic molecules that have not yet been totally separated and identified. However, single classes of residual biomolecules have been identified by selective extractions and chemical fractionation, and are represented by alkanoic and hydroxy alkanoic acids, resins, waxes, lignin residues, sugars, and peptides.

21.4 Health issues

Humic and fulvic acids, when present in treated drinking water, can react with the chemicals used in the chlorination process to form disinfection byproducts such as dihaloacetonitriles, which are toxic to humans.*[10]*[11]

21.5 Ecological effects

Organic matter soil amendments have been known by farmers to be beneficial to plant growth for longer than recorded history.*[12] However, the chemistry and function of the organic matter have been a subject of controversy since humans began their postulating about it in the 18th century. Until the time of Liebig, it was supposed that humus was used directly by plants, but, after Liebig had shown that plant growth depends upon inorganic compounds, many soil scientists held the view that organic matter was useful for fertility only as it was broken down with the release of its constituent nutrient elements into inorganic forms. At the present time, soil scientists hold a more holistic view and at least recognize that humus influences soil fertility through its effect on the water-holding capacity of the soil. Also, since plants have been shown to absorb and translocate the complex organic molecules of systemic insecticides, they can no longer discredit the idea that plants may be able to absorb the soluble forms of humus;*[13] this may in fact be an essential process for the uptake of otherwise insoluble iron oxides.

A study on the effects of Humic acid on plant growth was conducted at Ohio State University which said in part "humic acids increased plant growth" and that there were "relatively large responses at low application rates" *[14]

A 1998 study by scientists at the North Carolina State University College of Agriculture and Life Sciences showed that addition of humate to soil significantly increased root mass in creeping bentgrass turf.*[15]*[16]

21.6 Ancient masonry

In Ancient Egypt, according to archeology, straw was mixed with mud in order to produce building bricks. Straw produces stronger bricks that are less likely to break or lose their shape. Modern investigations have found that humic acid is released from straw when mixed with mud, basically a mixture of sand and clay. Humic acid increases clay's plasticity.*[17]

21.7 See also

- glomalin

- Potassium humate

21.8 References

[1] Stevenson F.J. (1994). *Humus Chemistry: Genesis, Composition, Reactions*. New York: John Wiley & Sons.

[2] Yamauchi, Masashige; Katayama, Sadamu; Todoroki, Toshiharu; Watanable, Toshio (1984). "Total synthesis of fulvic acid" . *Journal of the Chemical Society, Chemical Communications* (23): 1565–6. doi:10.1039/C39840001565. Synthesis of fulvic

acid (1a) was accomplished by a route involving selective ozonization of 9-propenylpyranobenzopyran (1c), obtained by a regioselective cyclization of the 2-methylsulphinylmethyl 1,3-dione(3c)

[3] Stevenson, F.J. (1994). *Humus Chemistry: Genesis, Composition, Reactions*, Wiley & Sons, New York, 1994, pp. 188-210, .

[4] Tan, K. H. 2014. *Humic matter in soil and the environment: principles and controversies*. 2nd ed. Boca Ranton: CRC Press, .

[5] Baigorri R; Fuentes M; González-Gaitano G; García-Mina JM; Almendros G; González-Vila FJ. (2009). "Complementary Multianalytical Approach To Study the Distinctive Structural Features of the Main Humic Fractions in Solution: Gray Humic Acid, Brown Humic Acid, and Fulvic Acid". *J Agric Food Chem*. **57** (8): 3266–72. doi:10.1021/jf8035353. PMID 19281175.

[6] Fiorentino G., Spaccini R., Piccolo A (2006). "Separation of molecular constituents from a humic acid by solid-phase extraction following a transesterification reaction". *Talanta* **68** (4): 1135–1142. doi:10.1016/j.talanta.2005.07.037. PMID 18970442.

[7] Ghabbour, E.A.; Davies, G. (Editors) (2001). *Humic Substances: Structures, Models and Functions*. Cambridge, U.K.: RSC publishing. ISBN 978-0-85404-811-3.

[8] Piccolo, A. (2002). "The Supramolecular structure of humic substances. A novel understanding of humus chemistry and implications in soil science". *Advances in Agronomy*. Advances in Agronomy **75**: 57–134. doi:10.1016/S0065-2113(02)75003-7. ISBN 978-0-12-000793-6.

[9] Tipping, E (1994). "'WHAM – a chemical equilibrium model and computer code for waters, sediments, and soils incorporating a discrete site/electrostatic model of ion-binding by humic substances". *Computers and Geosciences* **20** (6): 973–1023. Bibcode:1994CG.....20..973T. doi:10.1016/0098-3004(94)90038-8.

[10] Oliver, Barry G. (1983). "Dihaloacetonitriles in drinking water: Algae and fulvic acid as precursors". *Environmental Science & Technology* **17** (2): 80. Bibcode:1983EnST...17...80O. doi:10.1021/es00108a003.

[11] Peters, Ruud J.B.; De Leer, Ed W.B.; De Galan, Leo (1990). "Dihaloacetonitriles in Dutch drinking waters". *Water Research* **24** (6): 797. doi:10.1016/0043-1354(90)90038-8.

[12] Lapedes, Daniel N., ed. (1966). *McGraw-Hill encyclopedia of science and technology: an international reference work, Volume 12*. McGraw-Hill. p. 428. ISBN 0070452652. The value of adding organic matter to the soil in the form of animal manures, green manures, and crop residues for producing favorable soil tilth has been known since ancient times

[13] Pan American Union. Dept. of Cultural Affairs. División de Fomento Científico, Pan American Union. Dept. of Scientific Affairs, Organization of American States. Dept. of Scientific Affairs (1984). *Ciencia interamericana: Volumes 24–27*. And since plants have shown their ability to absorb and translocate the complex molecules of systemic insecticides, they can no longer discredit the idea that plants are able to absorb the soluble humic nutrients, containing by far ...

[14] Arancon, Norman Q.; Edwards, Clive. A.; Lee, Stephen; Byrne, Robert (2006). "Effects of humic acids from vermicomposts on plant growth" (PDF). *European Journal of Soil Biology* **42**: S65. doi:10.1016/j.ejsobi.2006.06.004.

[15] Cooper, R. J.; Liu, Chunhua; Fisher, D. S. (1998). "Influence of Humic Substances on Rooting and Nutrient Content of Creeping Bentgrass". *Crop Science* **38** (6): 1639. doi:10.2135/cropsci1998.0011183X003800060037x.

[16] Liu, Chunhua; Cooper, R. J. (August 1999). "Humic Substances Their Influence on Creeping Bentgrass Growth and Stress Tolerance" (PDF). *TurfGrass Trends*: 6.

[17] Lucas, A.; Harris, J.R. (1998). *Ancient Egyptian Materials and Industries*. New York: Dover Publications. p. 49. ISBN 0-486-40446-3.

21.9   Further reading

- Hessen, D.O.; Tranvik, L.J. (Editors) (1998). *Aquatic humic substances: ecology and biogeochemistry*. Berlin: Springer. ISBN 3-540-63910-1.

- Sillanpää, M. (Ed.) Natural Organic Matter in Water, Characterization and Treatment Methods ISBN 9780128015032

Chapter 22

Humus

This article is about the organic matter in soil. For the band, see Humus (band). For the food, see Hummus. For programming language, see Humus (programming language).

In soil science, **humus** (coined 1790–1800; from the Latin *humus*: earth, ground[1]) refers to the fraction of soil organic matter that is amorphous and without the "cellular structure characteristic of plants, micro-organisms or animals." [2] Humus significantly influences the bulk density of soil and contributes to moisture and nutrient retention. Soil formation begins with the weathering of humus. In agriculture, humus is sometimes also used to describe mature, or natural compost extracted from a forest or other spontaneous source for use to amend soil.[3] It is also used to describe a topsoil horizon that contains organic matter (humus type,[4] humus form,[5] humus profile).[6]

22.1 Humification

22.1.1 Transformation of organic matter into humus

The process of "humification" can occur naturally in soil, or in the production of compost. The importance of chemically stable humus is thought by some to be the fertility it provides to soils in both a physical and chemical sense,[7] though some agricultural experts put a greater focus on other features of it, such as its ability to suppress disease.[8] It helps the soil retain moisture[9] by increasing microporosity,[10] and encourages the formation of good soil structure.[11][12] The incorporation of oxygen into large organic molecular assemblages generates many active, negatively charged sites that bind to positively charged ions (cations) of plant nutrients, making them more available to the plant by way of ion exchange.[13] Humus allows soil organisms to feed and reproduce, and is often described as the "life-force" of the soil.[14][15]

It is difficult to define humus precisely; it is a highly complex substance, which is still not fully understood. Humus should be differentiated from decomposing organic matter. The latter is rough-looking material and remains of the original plant are still visible. Fully humified organic matter, on the other hand, has a uniform dark, spongy, jelly-like appearance, and is amorphous. It may remain like this for millennia or more.[16] It has no determinate shape, structure or character. However, humified organic matter, when examined under the microscope may reveal tiny plant, animal or microbial remains that have been mechanically, but not chemically, degraded.[17] This suggests a fuzzy boundary between humus and organic matter. In most literature, humus is considered an integral part of soil organic matter.[18]

Plant remains (including those that passed through an animal gut and were excreted as feces) contain organic compounds: sugars, starches, proteins, carbohydrates, lignins, waxes, resins, and organic acids. The process of organic matter decay in the soil begins with the decomposition of sugars and starches from carbohydrates, which break down easily as detritivores initially invade the dead plant organs, while the remaining cellulose and lignin break down more slowly.[19] Simple proteins, organic acids, starches and sugars break down rapidly, while crude proteins, fats, waxes and resins remain relatively unchanged for longer periods of time. Lignin, which is quickly transformed by white-rot fungi,[20] is one of the main precursors of humus,[21] together with by-products of microbial[22] and animal[23] activity. The end-

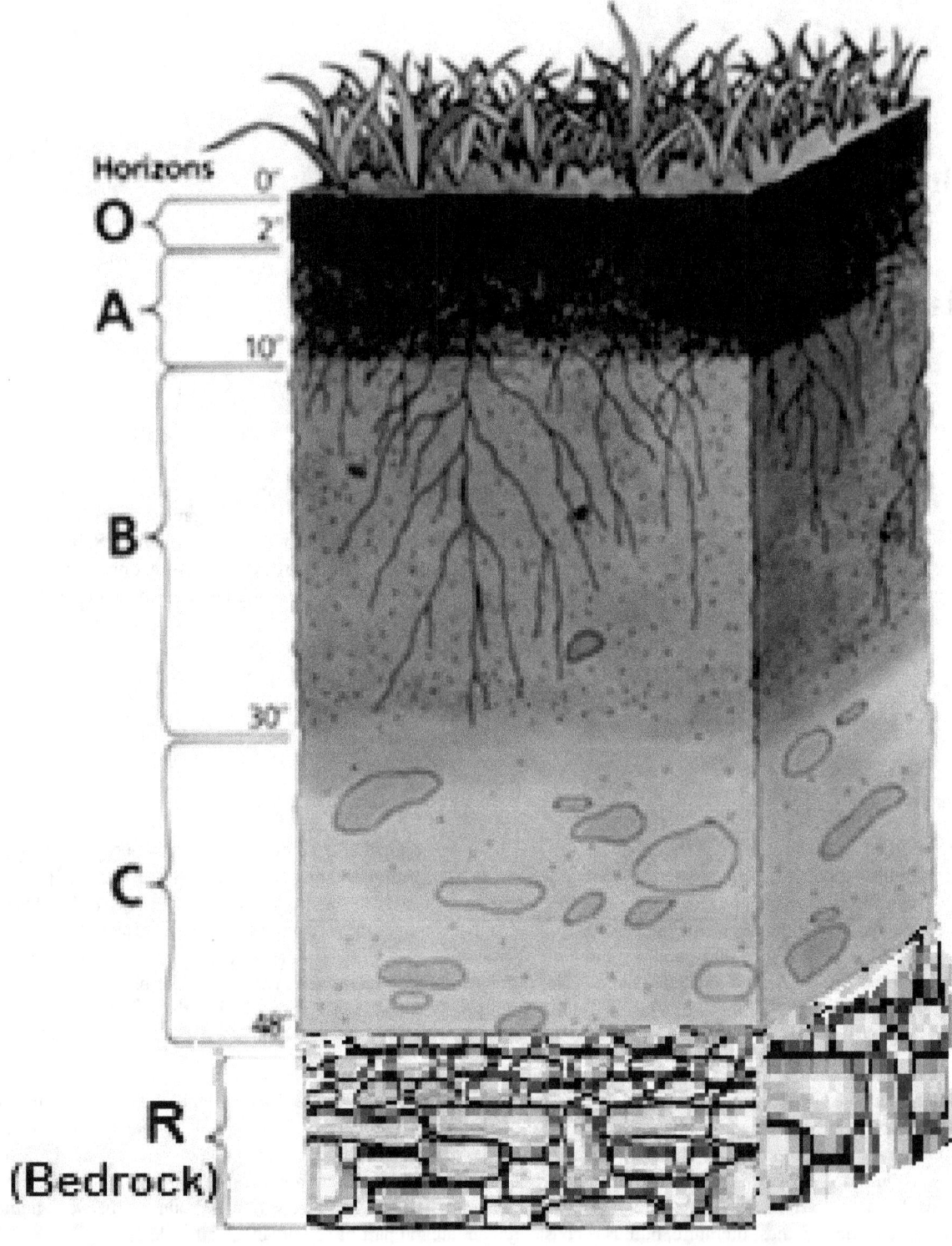

Horizons

O

A

B

C

R
(Bedrock)

0"
2"
10"
30"
48"

Humus has a characteristic black or dark brown color and is organic due to an accumulation of organic carbon. Soil scientists use the capital letters O, A, B, C, and E to identify the master horizons, and lowercase letters for distinctions of these horizons. Most soils have three major horizons—the surface horizon (A), the subsoil (B), and the substratum (C). Some soils have an organic horizon (O) on the surface, but this horizon can also be buried. The master horizon, E, is used for subsurface horizons that have a significant loss of minerals (eluviation). Hard bedrock, which is not soil, uses the letter R.

product of this process, the humus, is thus a mixture of compounds and complex life chemicals of plant, animal, or microbial origin that has many functions and benefits in the soil. Earthworm humus (vermicompost) is considered by some to be the best organic manure there is.*[24]

22.1.2 Stability

Much of the humus in most soils has persisted for more than a hundred years (rather than having been decomposed to CO_2), and can be regarded as stable; this is organic matter that has been protected from decomposition by microbial or enzyme action because it is hidden (occluded) inside small aggregates of soil particles or tightly attached (sorbed or complexed) to clays.*[25] Most humus that is not protected in this way is decomposed within ten years and can be regarded as less stable or more labile. Thus stable humus contributes little to the pool of plant-available nutrients in the soil, but it does play a part in maintaining its physical structure.*[26] A very stable form of humus is that formed from the slow oxidation of black carbon, after the incorporation of finely powdered charcoal into the topsoil. This process is thought to have been important in the formation of the fertile Amazonian dark earths or Terra preta do Indio.*[27]

22.2 Benefits of soil organic matter and humus

- The process that converts raw organic matter into humus feeds the soil population of microorganisms and other creatures, thus maintains high and healthy levels of soil life.*[14]*[15]

- The rate at which raw organic matter is converted into humus promotes (when fast) or limits (when slow) the coexistence of plants, animals, and microbes in soil.

- Effective humus and stable humus are further sources of nutrients to microbes, the former provides a readily available supply, and the latter acts as a longer-term storage reservoir.

- Decomposition of dead plant material causes complex organic compounds to be slowly oxidized (lignin-like humus) or to break down into simpler forms (sugars and amino sugars, aliphatic, and phenolic organic acids), which are further transformed into microbial biomass (microbial humus) or are reorganized, and further oxidized, into humic assemblages (fulvic and humic acids), which bind to clay minerals and metal hydroxides. There has been a long debate about the ability of plants to uptake humic substances from their root systems and to metabolize them. There is now a consensus about how humus plays a hormonal role rather than simply a nutritional role in plant physiology.*[28]*[29]

- Humus is a colloidal substance, and increases the soil's cation exchange capacity, hence its ability to store nutrients by chelation. While these nutrient cations are accessible to plants, they are held in the soil safe from being leached by rain or irrigation.*[13]

- Humus can hold the equivalent of 80–90% of its weight in moisture, and therefore increases the soil's capacity to withstand drought conditions.*[30]*[31]

- The biochemical structure of humus enables it to moderate – or buffer – excessive acid or alkaline soil conditions.*[32]

- During the humification process, microbes secrete sticky gum-like mucilages; these contribute to the crumb structure (tilth) of the soil by holding particles together, and allowing greater aeration of the soil.*[33] Toxic substances such as heavy metals, as well as excess nutrients, can be chelated (that is, bound to the complex organic molecules of humus) and so prevented from entering the wider ecosystem.*[34]

- The dark color of humus (usually black or dark brown) helps to warm up cold soils in the spring.

22.3   See also

- Soil Profile

- Biomass

- Biochar

- Terra preta

- Biotic material

- Soil horizon

- Detritus

- Glomalin

- Humic acid

- Organic matter

- Plant litter

- Mycorrhizal fungi and soil carbon storage

- Compost

22.4   References

[1] "humus." Dictionary.com Unabridged (v 1.1). Random House, Inc. 23 Sep 2008. Dictionary.com http://dictionary.reference.com/browse/humus.

[2] Whitehead, D. C.; Tinsley, J. (1963). "The biochemistry of humus formation". *Journal of the Science of Food and Agriculture* **14** (12): 849–857. doi:10.1002/jsfa.2740141201. Retrieved 26 July 2014.

[3] "humus." Encyclopædia Britannica. Encyclopædia Britannica Online. Encyclopædia Britannica Inc., 2011. Web. 24 Nov 2011. <http://www.britannica.com/EBchecked/topic/276408/humus>.

[4] Chertov, O.G.; Kornarov, A.S.; Crocker, G.; Grace, P.; Klir, J.; Körschens, M.; Poulton, P.R.; Richter, D. (1997). "Simulating trends of soil organic carbon in seven long-term experiments using the SOMM model of the humus types". *Geoderma* **81**: 121–135. doi:10.1016/S0016-7061(97)00085-2.

[5] Baritz, R., 2003. Humus forms in forests of the northern German lowlands. Schweizerbart, Stuttgart, Germany, 145 pp.

[6] Bunting, B.T.; Lundberg, J. (1995). "The humus profile-concept, class and reality". *Geoderma* **40**: 17–36. doi:10.1016/0016-7061(87)90011-5.

[7] Hargitai, L (1993). "The soil of organic matter content and humus quality in the maintenance of soil fertility and in environmental protection". *Landscape and Urban Planning* **27** (2–4): 161–167. doi:10.1016/0169-2046(93)90044-E.

[8] Hoitink, H.A.; Fahy, P.C. (1986). "Basic for the control of soilborne plant pathogens with composts". *Annual Review of Phytopathology* **24**: 93–114. doi:10.1146/annurev.py.24.090186.000521.

[9] C.Michael Hogan. 2010. *Abiotic factor*. Encyclopedia of Earth. eds Emily Monosson and C. Cleveland. National Council for Science and the Environment. Washington DC

[10] De Macedo, J.R.; Do Amaral, Meneguelli; Ottoni, T.B.; Araujo, Jorge Araújo; de Sousa Lima, J. (2002). "Estimation of field capacity and moisture retention based on regression analysis involving chemical and physical properties in Alfisols and Ultisols of the state of Rio de Janeiro". *Communications in Soil Science and Plant Analysis* **33** (13–14): 2037–2055. doi:10.1081/CSS-120005747.

[11] Hempfling, R.; Schulten, H.R.; Horn, R. (1990). "Relevance of humus composition to the physical/mechanical stability of agricultural soils: a study by direct pyrolysis-mass spectrometry". *Journal of Analytical and Applied Pyrolysis* **17** (3): 275–281. doi:10.1016/0165-2370(90)85016-G.

[12] Soil Development: Soil Properties

[13] Szalay, A (1964). "Cation exchange properties of humic acids and their importance in the geochemical enrichment of UO2++ and other cations". *Geochimica et Cosmochimica Acta* **28**: 1605–1614. doi:10.1016/0016-7037(64)90009-2.

[14] Elo, S.; Maunuksela, L.; Salkinoja-Salonen, M.; Smolander, A.; Haahtela, K. (2006). "Humus bacteria of Norway spruce stands: plant growth promoting properties and birch, red fescue and alder colonizing capacity". *FEMS Microbiology Ecology* **31**: 143–152. doi:10.1111/j.1574-6941.2000.tb00679.x.

[15] Vreeken-Buijs, M.J.; Hassink, J.; Brussaard, L. (1998). "Relationships of soil microarthropod biomass with organic matter and pore size distribution in soils under different land use". *Soil Biology and Biochemistry* **30**: 97–106. doi:10.1016/S0038-0717(97)00064-3.

[16] Di Giovanni1, C.; Disnar, J.R.; Bichet, V.; Campy, M. (1998). "Sur la présence de matières organiques mésocénozoïques dans des humus actuels (bassin de Chaillexon, Doubs, France)". *Comptes Rendus de l'Académie des Sciences de Paris, Series IIA, Earth and Planetary Science* **326**: 553–559. doi:10.1016/S1251-8050(98)80206-1.

[17] Nicolas Bernier and Jean-François Ponge (1994). "Humus form dynamics during the sylvogenetic cycle in a mountain spruce forest" (PDF). *Soil Biology and Biochemistry* **26** (2): 183–220. doi:10.1016/0038-0717(94)90161-9.

[18] Humintech® | Definition Of Soil Organic Matter & Humic Acids Based Products

[19] Berg, B., McClaugherty, C., 2007. Plant litter: decomposition, humus formation, carbon sequestration, 2nd ed. Springer, 338 pp., ISBN 3-540-74922-5

[20] Levin, L., Forchiassin, F., Ramos, A.M., 2002. Copper induction of lignin-modifying enzymes in the white-rot fungus Trametes trogii" *Mycologia* 94:377–383

[21] González-Pérez, M.; Vidal Torrado, P.; Colnago, L.A.; Martin-Neto, L.; Otero, X.L.; Milori, D.M.B.P.; Haenel Gomes, F. (2008). "13C NMR and FTIR spectroscopy characterization of humic acids in spodosols under tropical rain forest in southeastern Brazil". *Geoderma* **146**: 425–433. doi:10.1016/j.geoderma.2008.06.018.

[22] Knicker, H.; Almendros, G.; González-Vila, F.J.; Lüdemann, H.D.; Martin, F. (1995). "13C and 15N NMR analysis of some fungal melanins in comparison with soil organic matter". *Organic Geochemistry* **23**: 1023–1028. doi:10.1016/0146-6380(95)00094-1.

[23] Muscoloa, A.; Bovalob, F.; Gionfriddob, F.; Nardi, S. (1999). "Earthworm humic matter produces auxin-like effects on Daucus carota cell growth and nitrate metabolism". *Soil Biology and Biochemistry* **31**: 1303–1311. doi:10.1016/S0038-0717(99)00049-8.

[24] Vermiculture

[25] Dungait, J. A.; Hopkins, D. W.; Gregory, A. S.; Whitmore, A. P. (2012). "Soil organic matter turnover is governed by accessibility not recalcitrance" (PDF). *Global Change Biology* **18** (6): 1781–1796. doi:10.1111/j.1365-2486.2012.02665.x. Retrieved 30 August 2014.

[26] Oades, J. M. (1984). "Soil organic matter and structural stability: mechanisms and implications for management". *Plant and soil* **76**: 319–337. doi:10.1007/BF02205590. Retrieved 30 August 2014.

[27] Lehmann, J., Kern, D.C., Glaser, B., Woods, W.I., 2004. Amazonian Dark Earths: origin, properties, management. Springer, 523 pp. ISBN 978-1-4020-1839-8

[28] Eyheraguibel, B.; Silvestrea, J. Morard (2008). "Effects of humic substances derived from organic waste enhancement on the growth and mineral nutrition of maize". *Bioresource Technology* **99**: 4206–4212. doi:10.1016/j.biortech.2007.08.082.

[29] Zandonadi, D. B.; Santos, M. P.; Busato, J. G.; Peres, L. E. P.; Façanha, A. R. (2013). "Plant physiology as affected by humified organic matter". *Theoretical and Experimental Plant Physiology* **25**: 13–25. doi:10.1590/S2197-00252013000100003. Retrieved 30 August 2014.

[30] Olness, A.; Archer, D. (2005). "Effect of organic carbon on available water in soil". *Soil Science* **170**: 90–101. doi:10.1097/00010694-200502000-00002.

[31] Effect of Organic Carbon on Available Water in Soil : Soil Science

[32] Kikuchi, R (2004). "Deacidification effect of the litter layer on forest soil during snowmelt runoff: laboratory experiment and its basic formularization for simulation modeling". *Chemosphere* **54** (8): 1163–1169. doi:10.1016/j.chemosphere.2003.10.025. PMID 14664845.

[33] Caesar-Tonthat, T.C. (2002). "Soil binding properties of mucilage produced by a basidiomycete fungus in a model system". *Mycological Research* **106** (8): 930–937. doi:10.1017/S0953756202006330.

[34] Huang, D.L.; Zeng, G.M.; Feng, C.L.; Hu, S.; Jiang, X.Y.; Tang, L.; Su, F.F.; Zhang, Y.; Zeng, W.; Liu, H.L. (2008). "Degradation of lead-contaminated lignocellulosic waste by Phanerochaete chrysosporium and the reduction of lead toxicity". *Environmental Science and Technology* **42** (13): 4946–4951. Bibcode:2008EnST...42.4946H. doi:10.1021/es800072c. PMID 18678031.

22.5 External links

- Jerzy Weber. "Types of humus in soils". Agricultural University of Wroclaw, Poland. Retrieved 2013-12-12.

Chapter 23

John Innes compost

John Innes compost is a set of four formulae for growing medium, developed at the former John Innes Horticultural Institution (JIHI), now the John Innes Centre, in the 1930s and released into the public domain. The scientists who developed the formulae were William Lawrence and John Newell.*[1] The director at the time was Daniel Hall*[2] Lawrence started to investigate the whole procedure of making seed and potting composts following a major disaster in 1933 with *Primula sinensis* seedlings, an important experimental plant for JIHI geneticists. After hundreds of trials, Lawrence and Newell arrived at two basic composts, a base fertiliser for use in the potting compost and a standard feed. The formulae of these as yet unnamed composts were published in 1938.*[3] These composts originally provided a sterile and well balanced growing medium for the experimental plant material needed at the institute. The Institution made the formulae generally available, but never manufactured the composts for sale nor benefited financially from their production.*[4] The name 'John Innes Compost' was allotted in 1938–39; the horticultural retail trade in the composts made 'John Innes' a household name, but JIHI received no financial benefit from them.*[3] The formulae contain loam, peat, sand or grit, and fertiliser in varying ratios for specific purposes. The composts are "soil-based".

23.1 External links

- John Innes Centre information on John Innes Compost

- John Innes Manufacturers' Association

23.2 References

[1] History of the John Innes Compost

[2] A History of the John Innes Centre

[3] John Innes Centre historical timeline

[4] John Innes, more than just a compost

Chapter 24

Keyhole garden

A **keyhole garden** is a 2 meter wide circular raised garden with a keyhole-shaped indentation on one side. The indentation allows gardeners to add uncooked vegetable scraps, greywater, and manure into a composting basket that sits in the center of the bed. In this way, composting materials can be added to the basket throughout the growing season to provide nutrients for the plants. The upper layer of soil is hilled up against the center basket so the soil slopes gently down from the center to the sides. Most keyhole gardens rise about one meter above the ground and have walls made of stone. The stone wall not only gives the garden its form, but helps trap moisture within the bed. Keyhole gardens originated in Lesotho and are well adapted to dry arid lands and deserts. In Africa they are positioned close to the kitchen and used to raise leafy greens such as lettuce, kale, and spinach; herbs; and root crops such as onions, garlic, carrots, and beets. Keyhole gardens are ideal for intensive planting, a technique in which plants are placed close together to maximize production. Plants with wide reaching root systems such as tomatoes and zucchini may not perform well in a keyhole garden.

The keyhole garden was developed in Lesotho by the Consortium for Southern Africa Food Security Emergency (C-SAFE), based upon a design that originated with CARE in Zimbabwe. In the mid-1990's Lesotho had one of the highest AIDS/HIV rates in the world. C-SAFE designed the keyhole garden for people who suffered from AIDS or were otherwise unable to tend a traditional garden. They are tall enough that people don't have to bend over while working in them; sturdy enough that a person who is weak can lean against them while they work; and are small enough that the entire bed is within arm's reach. The garden is constructed using layers of compost, manure, wood ash and other nutrient rich materials so they are more productive than most home gardens; and they hold water making them drought resistant. The walls can be made of common stones picked up from a field, cinderblocks, bricks, or any material strong enough to hold in the soil. Clean water is used when watering the plants on the surface, while household greywater is poured down into the compost basket.

While they are designed for people who were too sick to tend a traditional garden, because they were so productive, people in good health started building them as kitchen gardens. Here the family could grow its high value crops using succession planting. In the end C-SAFE helped build over 20,000 keyhole gardens, and when they returned a couple of years later, more than 90% were still in use. Today keyhole gardens are found in many places though out Africa, including Ethiopia, Rwanda, Kenya, Sudan, and Nigeria.

African style keyhole gardens have been built in Texas and other places in the US. The Texas Master Gardeners Association has put on several workshops to promote them. They have modified the basic design, having standardized on a 6-foot wide bed with a 12-inch tube made of rabbit fencing or chicken wire for the compost basket. It is common in Texas to add red-wiggler worms to the compost basket to help break down the organic matter.

The term "keyhole garden" is also used in permaculture to describe a raised bed garden with a longer keyhole shaped path cut into it. Permaculture keyhole gardens tend to be 12"–18" high, much larger and flatter than the African variety, and do not incorporate a compost basket or compost pile built into the bed.

24.1 See also

- Raised-bed gardening
- Square foot gardening

Chapter 25

Leaf mold

Leaf mold is a form of compost produced by the fungal breakdown[*][1] of shrub and tree leaves, which are generally too dry, acidic, or low in nitrogen for bacterial decomposition.

25.1 Description

Due to the slow decaying nature of their high carbon content,[*][2] dry leaves break down far more slowly than most other compost ingredients. This can be overcome either by placing the collected leaves wet in plastic bags (taking care to avoid collecting from areas that may be subject to high levels of pollution, e.g., roadsides), or in specially constructed wire bins, to encourage fungal action. To accelerate this fungal breakdown, it is useful to keep the leaves wet and avoid the drying effects of wind. The traditional wire enclosure may slow down the process by allowing the contents to dry out unless it is lined with cardboard or similar material. 1/

25.2 Time and process

Leaves alone can take between one and two years to break down into rich humic matter with a smell reminiscent of ancient woodland. While not high in nutrient content, leaf mold is an excellent humic soil conditioner. To speed up the decomposition process, fallen leaves can be shredded, for instance by using a rotary lawn mower. Adding fresh grass clippings[*][3] to autumn leaves will also speed the process. For best results watch the pile to keep moisture content high enough, observe temperatures, and turn the pile occasionally to improve the cycle.

25.3 See also

- Worm compost
- Spent mushroom compost
- Recycling

25.4 References

[1] Compost organisms

[2] Compost Chemistry

[3] Composting Ingredients

25.5 External links

- BBC Gardening How to Make Leaf Mould

- Green Fingers How to Make Leaf Mould

- Leaves & Leaf Mold, nature's mulch & top-coating at The Garden of Paghat

Chapter 26

Mulch

A **mulch** is a layer of material applied to the surface of an area of soil. Its purpose is any or all of the following:

- to conserve moisture

- to improve the fertility and health of the soil

- to reduce weed growth

- to enhance the visual appeal of the area

A mulch is usually but not exclusively organic in nature. It may be permanent (e.g. plastic sheeting) or temporary (e.g. bark chips). It may be applied to bare soil, or around existing plants. Mulches of manure or compost will be incorporated naturally into the soil by the activity of worms and other organisms. The process is used both in commercial crop production and in gardening, and when applied correctly can dramatically improve soil productivity.*[1]

26.1   Uses

Many materials are used as mulches, which are used to retain soil moisture, regulate soil temperature, suppress weed growth, and for aesthetics.*[2] They are applied to the soil surface,*[3] around trees, paths, flower beds, to prevent soil erosion on slopes, and in production areas for flower and vegetable crops. Mulch layers are normally two inches or more deep when applied.*[4]*[5]

They are applied at various times of the year depending on the purpose. Towards the beginning of the growing season mulches serve initially to warm the soil by helping it retain heat which is lost during the night. This allows early seeding and transplanting of certain crops, and encourages faster growth. As the season progresses, mulch stabilizes the soil temperature and moisture, and prevents the growing of weeds from seeds.*[6] In temperate climates, the effect of mulch is dependent upon the time of year they are applied and when applied in fall and winter, are used to delay the growth of perennial plants in the spring or prevent growth in winter during warm spells, which limits freeze thaw damage.*[7]

The effect of mulch upon soil moisture content is complex. Mulch forms a layer between the soil and the atmosphere which prevents sunlight from reaching the soil surface, thus reducing evaporation. However, mulch can also prevent water from reaching the soil by absorbing or blocking water from light rains.

In order to maximise the benefits of mulch, while minimizing its negative influences, it is often applied in late spring/early summer when soil temperatures have risen sufficiently, but soil moisture content is still relatively high.*[8] However, permanent mulch is also widely used and valued for its simplicity, as popularized by author Ruth Stout, who said, "My way is simply to keep a thick mulch of any vegetable matter that rots on both sides of my vegetable and flower garden all year long. As it decays and enriches the soils, I add more." *[9]

Plastic mulch used in large-scale commercial production is laid down with a tractor-drawn or standalone layer of plastic mulch. This is usually part of a sophisticated mechanical process, where raised beds are formed, plastic is rolled out on top, and seedlings are transplanted through it. Drip irrigation is often required, with drip tape laid under the plastic, as plastic mulch is impermeable to water.

26.2 Materials

Materials used as mulches vary and depend on a number of factors. Use takes into consideration availability, cost, appearance, the effect it has on the soil—including chemical reactions and pH, durability, combustibility, rate of decomposition, how clean it is—some can contain weed seeds or plant pathogens.[6]

A variety of materials are used as mulch:

- Organic residues: grass clippings, leaves, hay, straw, kitchen scraps comfrey, shredded bark, whole bark nuggets, sawdust, shells, woodchips, shredded newspaper, cardboard, wool, animal manure, etc. Many of these materials also act as a direct composting system, such as the mulched clippings of a mulching lawn mower, or other organics applied as sheet composting.

- Compost: fully composted materials are used to avoid possible phytotoxicity problems. Materials that are free of seeds are ideally used, to prevent weeds introduced by the mulch.

- Rubber mulch: made from recycled tire rubber.

- Plastic mulch: crops grow through slits or holes in thin plastic sheeting. This method is predominant in large-scale vegetable growing, with millions of acres cultivated under plastic mulch worldwide each year (disposal of plastic mulch is cited as an environmental problem).

- Rock and gravel can also be used as a mulch. In cooler climates the heat retained by rocks may extend the growing season.

In some areas of the United States, such as central Pennsylvania and northern California, mulch is often referred to as "tanbark", even by manufacturers and distributors. In these areas, the word "mulch" is used specifically to refer to very fine tanbark or peat moss.

26.2.1 Organic mulches

Organic mulches decay over time and are temporary. The way a particular organic mulch decomposes and reacts to wetting by rain and dew affects its usefulness.

Some mulches such as straw, peat, sawdust and other wood products may for a while negatively affect plant growth because of their wide carbon to nitrogen ratio,[10] because bacteria and fungi that decompose the materials remove nitrogen from the surrounding soil for growth.[11][12] However, whether this effect has any practical impact on gardens is disputed by researchers and the experience of gardeners.[13] Organic mulches can mat down, forming a barrier that blocks water and air flow between the soil and the atmosphere. Vertically applied organic mulches can wick water from the soil to the surface, which can dry out the soil.[14] Mulch made with wood can contain or feed termites, so care must be taken about not placing mulch too close to houses or building that can be damaged by those insects. Some mulch manufacturers recommend putting mulch several inches away from buildings.

Commonly available organic mulches include:[6]

Leaves

- *Leaves* from deciduous trees, which drop their foliage in the autumn/fall. They tend to be dry and blow around in the wind, so are often chopped or shredded before application. As they decompose they adhere to each other but also allow water and moisture to seep down to the soil surface. Thick layers of entire leaves, especially of maples and oaks, can form a soggy mat in winter and spring which can impede the new growth lawn grass and other plants. Dry leaves are used as winter mulches to protect plants from freezing and thawing in areas with cold winters, they are normally removed during spring.

Grass clippings

- *Grass clippings*, from mowed lawns are sometimes collected and used elsewhere as mulch. Grass clippings are dense and tend to mat down, so are mixed with tree leaves or rough compost to provide aeration and to facilitate their decomposition without smelly putrefaction. Rotting fresh grass clippings can damage plants; their rotting often produces a damaging buildup of trapped heat. Grass clippings are often dried thoroughly before application, which mediates against rapid decomposition and excessive heat generation. Fresh green grass clippings are relatively high in nitrate content, and when used as a mulch, much of the nitrate is returned to the soil, but the routine removal of grass clippings from the lawn results in nitrogen deficiency for the lawn.

Peat moss

- *Peat moss*, or sphagnum peat, is long lasting and packaged, making it convenient and popular as a mulch. When wetted and dried, it can form a dense crust that does not allow water to soak in. When dry it can also burn, producing a smoldering fire. It is sometimes mixed with pine needles to produce a mulch that is friable. It can also lower the pH of the soil surface, making it useful as a mulch under acid loving plants.

However peat bogs are a valuable wildlife habitat, and peat is also one of the largest stores of carbon (in Britain, out of a total estimated 9952 million tonnes of carbon in British vegetation and soils, 6948 million tonnes carbon are estimated to be in Scottish, mostly peatland, soils*[15]), so gardeners who wish to protect the environment will choose more sustainable alternatives.*[16]

Wood chips

- *Wood chips* are a byproduct of the pruning of trees by arborists, utilities and parks; they are used to dispose of bulky waste. Tree branches and large stems are rather coarse after chipping and tend to be used as a mulch at least three inches thick. The chips are used to conserve soil moisture, moderate soil temperature and suppress weed growth. The decay of freshly produced chips from recently living woody plants, consumes nitrate; this is often off set with a light application of a high-nitrate fertilizer. Wood chips are most often used under trees and shrubs. When used around soft stemmed plants, an unmulched zone is left around the plant stems to prevent stem rot or other possible diseases. They are often used to mulch trails, because they are readily produced with little additional cost outside of the normal disposal cost of tree maintenance. Wood chips come in various colors.

Woodchip mulch

- *Woodchip mulch* is a byproduct of reprocessing used (untreated) timber (usually packaging pallets), to dispose of wood waste by creating woodchip mulch. The chips are used to conserve soil moisture, moderate soil temperature and suppress weed growth. Woodchip mulch is often used under trees, shrubs or large planting areas and can last much longer than arborist mulch. In addition, many consider woodchip mulch to be visually appealing, as it comes in various colors. Woodchips can also be reprocessed into playground woodchip to be used as an impact-attenuating playground surfacing.

Bark chips

- *Bark chips* of various grades are produced from the outer corky bark layer of timber trees. Sizes vary from thin shredded strands to large coarse blocks. The finer types are very attractive but have a large exposed surface area

Bark chips

that leads to quicker decay. Layers two or three inches deep are usually used, bark is relativity inert and its decay does not demand soil nitrates. Bark chips are also available in various colors.

Straw mulch / field hay / salt hay

- *Straw mulch* or *field hay* or *salt hay* are lightweight and normally sold in compressed bales. They have an unkempt look and are used in vegetable gardens and as a winter covering. They are biodegradable and neutral in pH. They have good moisture retention and weed controlling properties but also are more likely to be contaminated with weed seeds. Salt hay is less likely to have weed seeds than field hay. Straw mulch is also available in various colors.

Cardboard / newspaper

- *Cardboard* or *newspaper* can be used as mulches. These are best used as a base layer upon which a heavier mulch such as compost is placed to prevent the lighter cardboard/newspaper layer from blowing away. By incorporating a layer of cardboard/newspaper into a mulch, the quantity of heavier mulch can be reduced, whilst improving the weed suppressant and moisture retaining properties of the mulch.*[8] However, additional labour is expended when planting through a mulch containing a cardboard/newspaper layer, as holes must be cut for each plant. Sowing seed through mulches containing a cardboard/newspaper layer is impractical. Application of newspaper mulch in windy weather can be facilitated by briefly pre-soaking the newspaper in water to increase its weight.

Permaculture garden with a fruit tree, herbs, flowers and vegetables mulched with hay

26.3   Colored Mulch

Some organic mulches are colored red, brown, black, and other colors. Isopropanolamine, specifically 1-Amino-2-propanol or DOW™ monoisopropanolamine, may be used as a pigment dispersant and color fastener in these mulches.*[17]*[18]*[19]*
Types of mulch which can be dyed include: wood chips, bark chips (barkdust) and pine straw. Colored mulch is made by dyeing the mulch in a water-based solution of colorant and chemical binder. When colored mulch first entered the market, most formulas were suspected to contain toxic, heavy metals and other contaminates. Today, "current investigations indicate that mulch colorants pose no threat to people, pets or the environment. The dyes currently used by the mulch and soil industry are similar to those used in the cosmetic and other manufacturing industries (i.e., iron oxide)," as stated by the Mulch and Soil Council.*[21] Colored mulch can be applied anywhere non-colored mulch is used (such as large bedded areas or around plants) and features many of the same gardening benefits as traditional mulch, such as improving soil productivity and retaining moisture.*[22] As mulch decomposes, just as with non-colored mulch, more mulch may need to be added to continue providing benefits to the soil and plants. However, if mulch is faded, spraying dye to previously spread mulch in order to restore color is an option.*[23]

26.4   Anaerobic (sour) mulch

Mulch normally smells like freshly cut wood, but sometimes develops a toxicity that causes it to smell like vinegar, ammonia, sulfur or silage. This happens when material with ample nitrogen content is not rotated often enough and it forms pockets of increased decomposition. When this occurs, the process may become anaerobic and produce these phytotoxic materials in small quantities. Once exposed to the air, the process quickly reverts to an aerobic process, but these toxic materials may be present for a period of time. If the mulch is placed around plants before the toxicity has had a chance to dissipate, then the plants could very likely be damaged or killed depending on their hardiness. Plants that are predominantly low to the ground or freshly planted are the most susceptible, and the phytotoxicity may prevent germination of some seeds.*[24]

If sour mulch is applied and there is plant kill, the best thing to do is to water the mulch heavily. Water dissipates the chemicals faster and refreshes the plants. Removing the offending mulch may have little effect, because by the time plant kill is noticed, most of the toxicity is already dissipated. While testing after plant kill will not likely turn up anything, a simple pH check may reveal high acidity, in the range of 3.8 to 5.6 instead of the normal range of 6.0 to 7.2. Finally, placing a bit of the offending mulch around another plant to check for plant kill will verify if the toxicity has departed. If the new plant is also killed, then sour mulch is probably not the problem.

26.5 Groundcovers (living mulches)

Main articles: Groundcovers and Living mulch

Groundcovers are plants which grow close to the ground, under the main crop, to slow the development of weeds and provide other benefits of mulch. They are usually fast-growing plants that continue growing with the main crops. By contrast, cover crops are incorporated into the soil or killed with herbicides. However, live mulches also may need to be mechanically or chemically killed eventually to prevent competition with the main crop.*[25]

Some groundcovers can perform additional roles in the garden such as nitrogen fixation in the case of clovers, dynamic accumulation of nutrients from the subsoil in the case of creeping comfrey (*Symphytum ibericum*), and even food production in the case of *Rubus tricolor*.*[26]

26.6 On-site mulch production

Owing to the great bulk of mulch which is often required on a site, it is often impractical and expensive to source and import sufficient mulch materials. An alternative to importing mulch materials is to grow them on site in a "mulch garden" - an area of the site dedicated entirely to the production of mulch which is then transferred to the growing area.*[26] Mulch gardens should be sited as close as possible to the growing area so as to facilitate transfer of mulch materials.*[26]

26.7 Mulching (composting) over unwanted plants

Main article: Sheet mulching

Sufficient mulch over plants will destroy them, and may be more advantageous than using herbicide, cutting, mowing, pulling, raking, or tilling. The higher the temperature that this "mulch" is composted, the quicker the reduction of undesirable materials. "Undesirable materials" may include living seed, plant "trash", as well as pathogens such as from animal feces, urine (e.g. hantavirus), fleas, lice, ticks, etc.

In some ways this improves the soil by attracting and feeding earthworms, and adding humus. Earthworms "till" the soil, and their feces are among the best fertilizers and soil conditioners.

Urine may be toxic to plants if applied to growing areas undiluted. See Compost ingredients: Urine.

26.8 Polypropylene and polyethylene mulch

Polypropylene mulch is made up of polypropylene polymers where polyethylene mulch is made up of polyethylene polymers. These mulches are commonly used in many plastics. Polyethylene is used mainly for weed reduction, where polypropylene is used mainly on perennials.[27] This mulch is placed on top of the soil and can be done by machine or hand with pegs to keep the mulch tight against the soil. This mulch can prevent soil erosion, reduce weeding, conserve soil moisture, and increase temperature of the soil.[28] Ultimately this can reduce the amount of work a farmer may have to do, and the amount of herbicides applied during the growing period. The black and clear mulches capture sunlight and warm the soil increasing the growth rate. White and other reflective colours will also warm the soil, but they do not suppress weeds as well.[28] This mulch may require other sources of obtaining water such as drip irrigation since it can reduce the amount of water that reaches the soil.[28] This mulch needs to be manually removed at the end of the season since when it starts to break down it breaks down into smaller pieces.[29] If the mulch is not removed before it starts to break down eventually it will break down into ketones and aldehydes polluting the soil.[29] This mulch is technically biodegradable but does not break down into the same materials the more natural biodegradable mulch does.

26.9 Biodegradable mulch

Quality biodegradable mulches are made out of plant starches and sugars or polyester fibres. These starches can come from plants such as wheat and corn.[30] These mulch films may be a bit more permeable allowing more water into the soil. This mulch can prevent soil erosion, reduce weeding, conserve soil moisture, and increase temperature of the soil.[28] Ultimately this can reduce the amount of herbicides used and manual labour farmers may have to do throughout the growing season. At the end of the season these mulches will start to break down from heat. Microorganisms in the soil break down the mulch into two components, water and CO_2, leaving no toxic residues behind.[30] This source of mulch is even less manual labour since it does not need to be removed at the end of the season and can actually be tilled into the soil.[30] With this mulch its important to take into consideration that its mulch is more delicate then other kinds. It should be placed on a day which is not too hot and with less tension then other synthetic mulches.[30] These also can be placed by machine or hand but its ideal to have a more starchy mulch that will allow it to stick to the soil better.

26.10 See also

- Forestry mulching

- Good Agricultural Practices

- Rubber mulch

- Plasticulture

- Integrated pest management

- Living mulch

26.11 References

[1] *RHS A-Z encyclopedia of garden plants.* United Kingdom: Dorling Kindersley. 2008. p. 1136. ISBN 1405332964.

[2] Alfred J. Turgeon; Lambert Blanchard McCarty; Nick Edward Christians (2009). *Weed control in turf and ornamentals.* Prentice Hall. p. 126. ISBN 978-0-13-159122-6.

[3] Mahesh K. Upadhyaya; Robert E. Blackshaw (2007). *Non-chemical Weed Management: Principles, Concepts and Technology.* CABI. pp. 135–. ISBN 978-1-84593-291-6.

[4] *Vegetable Gardening: Growing and Harvesting Vegetables.* Murdoch Books. 2004. pp. 110–. ISBN 978-1-74045-519-0.

[5] Dennis R. Pittenger (2002). *California Master Gardener Handbook.* UCANR Publications. pp. 567–. ISBN 978-1-879906-54-9.

[6] Louise; Bush-Brown, James (1996). *America's garden book*. New York: Macmillan USA. p. 768. ISBN 0-02-860995-6{{inconsistent citations}}

[7] Leon C. Snyder (2000). *Gardening in the Upper Midwest*. University of Minnesota Press. pp. 47–. ISBN 978-0-8166-3838-3.

[8] Patrick Whitefield, 2004, *The Earth Care Manual*, Permanent Publications, ISBN 978-1-85623-021-6

[9] Stout, Ruth. *Gardening Without Work*. Devon-Adair Press, 1961. Reprinted by Norton Creek Press, 2011, pp. 6-7. ISBN 978-0-9819284-6-3

[10] http://www.eau.ee/~{}agronomy/vol07Spec1/p7sI53.pdf

[11] http://joa.isa-arbor.com/request.asp?JournalID=1&ArticleID=3111&Type=2

[12] Jeff Gillman (1 February 2008). *The Truth About Organic Gardening: Benefits, Drawnbacks, and the Bottom Line*. Timber Press. pp. 51–. ISBN 978-1-60469-005-7.

[13] Stout, Ruth. *Gardening Without Work*. Devon-Adair Press, 1961. Reprinted by Norton Creek Press, 2011, pp. 192-193. ISBN 978-0-9819284-6-3

[14] David A. Bainbridge (11 June 2007). *A Guide for Desert and Dryland Restoration: New Hope for Arid Lands*. Island Press. pp. 239–. ISBN 978-1-61091-082-8.

[15] Milne, R.; T. A. Brown (1997). "Carbon in the vegetation and soils of Great Britain". *Journal of Environmental Management* **49**: 413–433. doi:10.1006/jema.1995.0118.

[16] Walker, John (2011). *How to Create an Eco Garden: The practical guide to greener, planet-friendly gardening*. Wigston, Leicestershire: Aquamarine. p. 33. ISBN 9781903141892.

[17] Product Information - DOW™ Monoisopropanolamine (MIPA)

[18] Product Safety Assessment - DOW™ Monoisopropanolamine

[19] 2010 Mulch Magic Red Material Safety Data Sheet

[20] 2007 Mulch Magic Red Material Safety Data Sheet

[21] http://www.mulchandsoilcouncil.org/faqs/mulch.php

[22] http://www.hort.purdue.edu/ext/mulch.html

[23] http://homeguides.sfgate.com/there-spray-can-use-renew-mulch-color-64513.html

[24] Beware of Sour Mulch

[25] Brandsaeter et al. 1998, Tharp and Kells, 2001

[26] Jacke and Toensmeier, Edible Forest Gardening, vol. II

[27] Dovorak, P. "BLACK POLYPROPYLENE MULCH TEXTILE IN ORGANIC AGRICULTURE" (PDF). *Czech University of Life Science Prague, Kamýcká* **52**. Retrieved 16 November 2014.

[28] Shonbeck, Dr. Mark (12 September 2012). "Synthetic Mulching Materials for Weed Management". *Extension*. Retrieved 16 November 2014.

[29] Corbin, A (2013). "Using Biodegradable Plastics as Agricultural Mulches." (PDF). Retrieved 16 November 2014.

[30] "Biodegradable Mulch" (PDF). *Penn State Extension*. Retrieved 16 November 2014.

26.12 External links

- Mulching Trees & Shrubs

Chapter 27

Multrum

A **multrum** is a large composting vessel, predominantly meant to decompose toilet excreta but also other organic residue. It is originally a composite word consisting of "multna" which means moldering or composting in Swedish and "rum" which is the Swedish word for room. A multrum has over several decades become a noun and has come to mean any large composting chamber connected to a toilet. This should not be confused with Clivus multrum which is a proprietary product. In Scandinavia there are many kinds of composting toilet multrums like Mullis, CompostEra besides Clivus Multrum.

Chapter 28

Nematode

This article is about the organism. For the infection, see Helminthiasis.

The **nematodes** /ˈnɛmətoʊdz/ or **roundworms** constitute the phylum **Nematoda**. They are a diverse animal phylum inhabiting a very broad range of environments. Nematode species can be difficult to distinguish, and although over 25,000 have been described,[2][3] of which more than half are parasitic, the total number of nematode species has been estimated to be about 1 million.[4] Unlike the phyla Cnidarians and Platyhelminthes (flatworms), nematodes have tubular digestive systems with openings at both ends.

Nematodes have successfully adapted to nearly every ecosystem from marine (salt water) to fresh water, to soils, and from the polar regions to the tropics, as well as the highest to the lowest of elevations. They are ubiquitous in freshwater, marine, and terrestrial environments, where they often outnumber other animals in both individual and species counts, and are found in locations as diverse as mountains, deserts and oceanic trenches. They are found in every part of the earth's lithosphere.[5] They represent 90% of all life forms on the ocean floor.[6] Their numerical dominance, often exceeding a million individuals per square meter and accounting for about 80% of all individual animals on earth, their diversity of life cycles, and their presence at various trophic levels point at an important role in many ecosystems.[7] Nematodes have even been found at great depth (0.9–3.6 km) below the surface of the Earth in gold mines in South Africa.[8][9][10][11][12]

Their many parasitic forms include pathogens in most plants and animals (including humans).[13] Some nematodes can undergo cryptobiosis. One group of carnivorous fungi, the nematophagous fungi, are predators of soil nematodes. They set enticements for the nematodes in the form of lassos or adhesive structures.[14][15][16]

Nathan Cobb the nematologist, described the ubiquity of nematodes on Earth thus:

> In short, if all the matter in the universe except the nematodes were swept away, our world would still be dimly recognizable, and if, as disembodied spirits, we could then investigate it, we should find its mountains, hills, vales, rivers, lakes, and oceans represented by a film of nematodes. The location of towns would be decipherable, since for every massing of human beings there would be a corresponding massing of certain nematodes. Trees would still stand in ghostly rows representing our streets and highways. The location of the various plants and animals would still be decipherable, and, had we sufficient knowledge, in many cases even their species could be determined by an examination of their erstwhile nematode parasites." [17]

28.1   Taxonomy and systematics

See also: List of nematode families

The group was originally defined by Karl Rudolphi in 1808[18] under the name **Nematoidea**, from Ancient Greek νῆμα (nêma, nêmatos, 'thread') and -ειδής (-eidēs, 'species'). It was reclassified as family **Nematodes** by Burmeister in 1837[18] and order **Nematoda** by K. M. Diesing in 1861.[18]

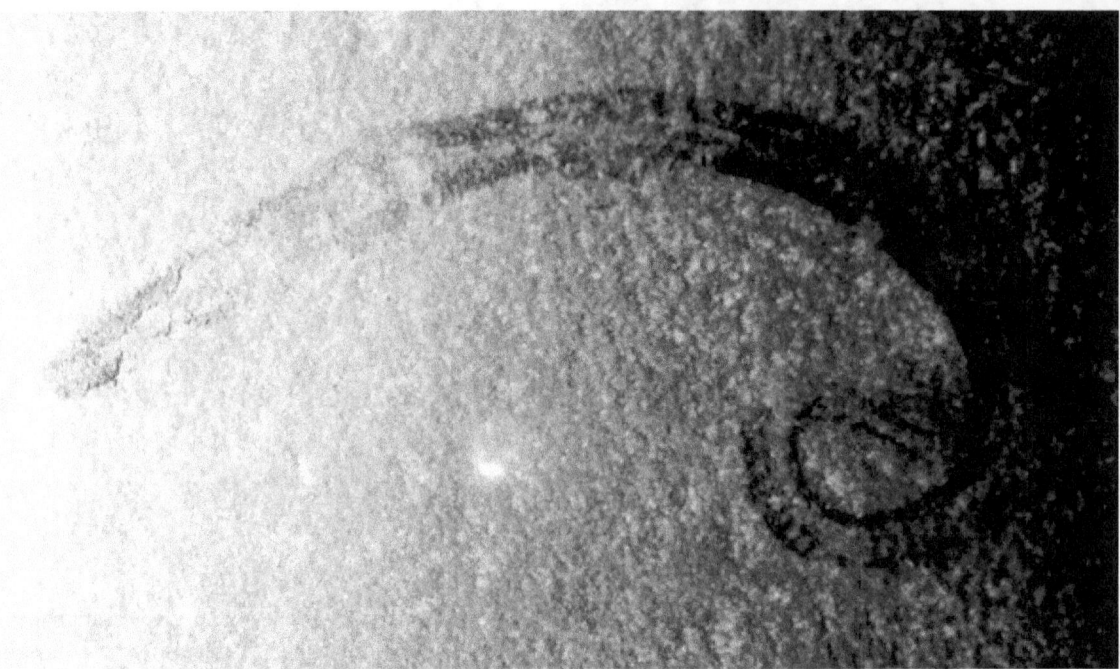

Eophasma jurasicum, *a fossilized nematode*

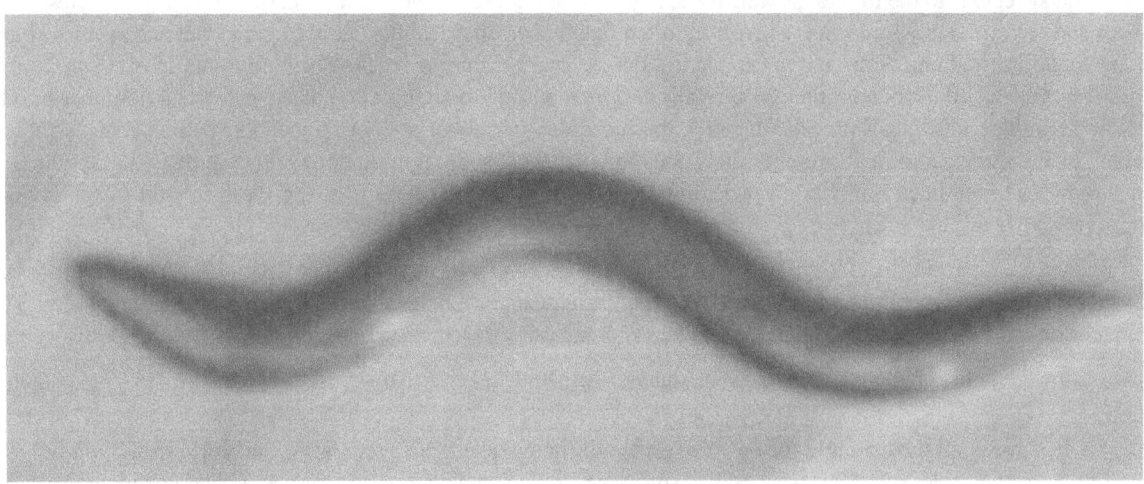

Caenorhabditis elegans

At its origin, the "Nematoidea" erroneously included Nematodes and Nematomorphs and Gordiacei, attributed by von Siebold (1843). Along with Acanthocephala, Trematoda and Cestoidea, it formed the group Entozoa.*[19] They were classed along with Acanthocephala in the new phylum Nemathelminthes (today obsolete) by Gegenbaur (1859). The taxon Nematoidea, including the family Gordiidae (horsehair worms), was then promoted to the rank of phylum by Ray Lankester (1877). In 1919, Nathan Cobb proposed that Nematode should be recognized alone as a phylum. He argued it should be called **nema** in English rather than "nematodes" *[lower-alpha 1] and defined the taxon **Nemates** (Latin plural of *nema*). Since Cobb was the first to exclude all but nematodes from the group, some sources consider the valid taxon name to be Nemates or Nemata, rather than Nematoda.*[20]

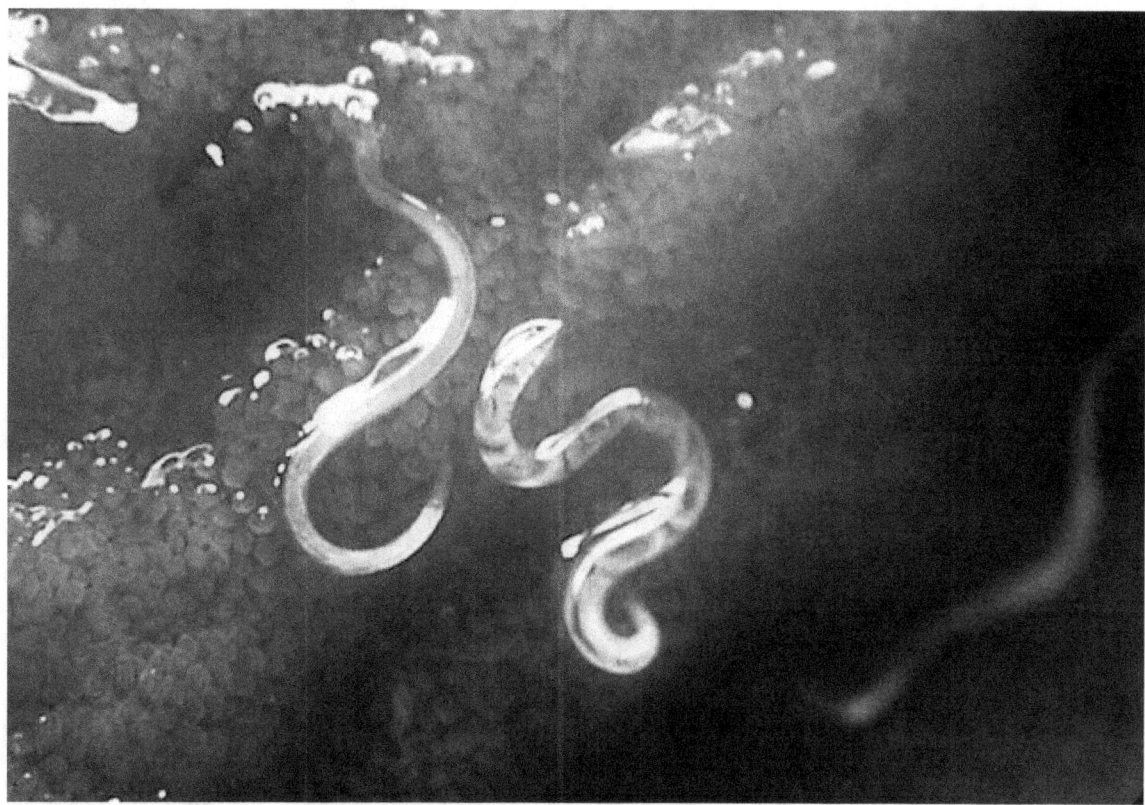

Rhabditia

28.1.1 Phylogeny

The phylogenetic relationships of the nematodes and their close relatives among the protostomian Metazoa are unresolved. Traditionally, they were held to be a lineage of their own but in the 1990s, they were proposed to form the group Ecdysozoa together with moulting animals, such as arthropods. The identity of the closest living relatives of the Nematoda has always been considered to be well resolved. Morphological characters and molecular phylogenies agree with placement of the roundworms as a sister taxon to the parasitic Nematomorpha; together they make up the Nematoida. Together with the Scalidophora (formerly Cephalorhyncha), the Nematoida form the Introverta. It is entirely unclear whether the Introverta are, in turn, the closest living relatives of the enigmatic Gastrotricha; if so, they are considered a clade Cycloneuralia, but much disagreement occurs both between and among the available morphological and molecular data. The Cycloneuralia or the Introverta—depending on the validity of the former—are often ranked as a superphylum.*[21]

28.1.2 Nematode systematics

Due to the lack of knowledge regarding many nematodes, their systematics is contentious. An earliest and influential classification was proposed by Chitwood and Chitwood*[22]—later revised by Chitwood*[23]—who divided the phylum into two—the Aphasmidia and the Phasmidia. These were later renamed Adenophorea (gland bearers) and Secernentea (secretors), respectively.*[24] The Secernentea share several characteristics, including the presence of phasmids, a pair of sensory organs located in the lateral posterior region, and this was used as the basis for this division. This scheme was adhered to in many later classifications, though the Adenophorea were not a uniform group.

Initial DNA sequence studies suggested the existence of five clades:*[25]

- Dorylaimia

- Enoplia

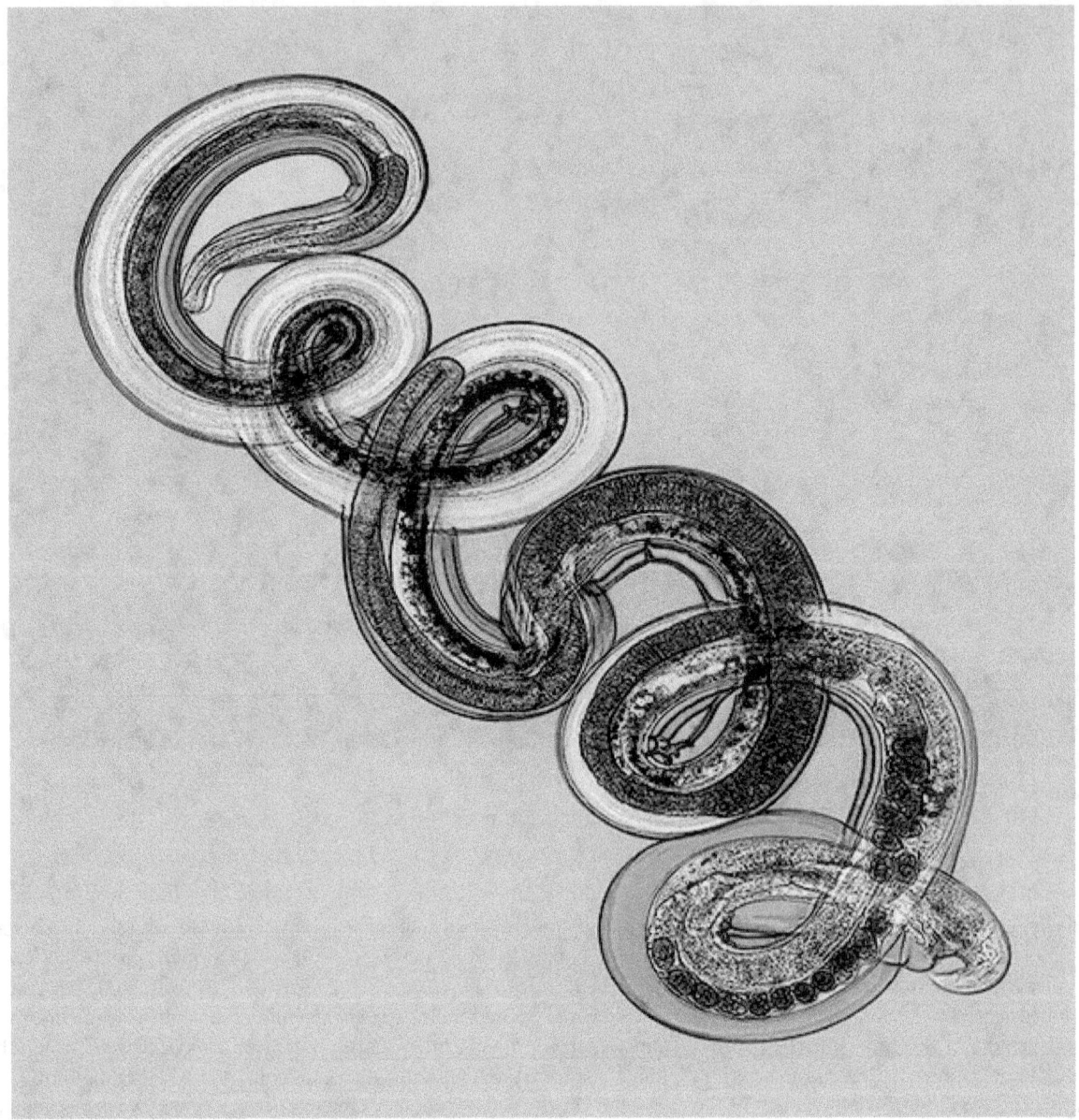

Nippostrongylus brasiliensis

- Spirurina

- Tylenchina

- Rhabditina

As it seems, the Secernentea are indeed a natural group of closest relatives. But the "Adenophorea" appear to be a paraphyletic assemblage of roundworms simply retaining a good number of ancestral traits. The old Enoplia do not seem to be monophyletic either, but to contain two distinct lineages. The old group "Chromadoria" seem to be another paraphyletic assemblage, with the Monhysterida representing a very ancient minor group of nematodes. Among the Secernentea, the Diplogasteria may need to be united with the Rhabditia, while the Tylenchia might be paraphyletic with the Rhabditia.[*][26]

The understanding of roundworm systematics and phylogeny as of 2002 is summarised below:

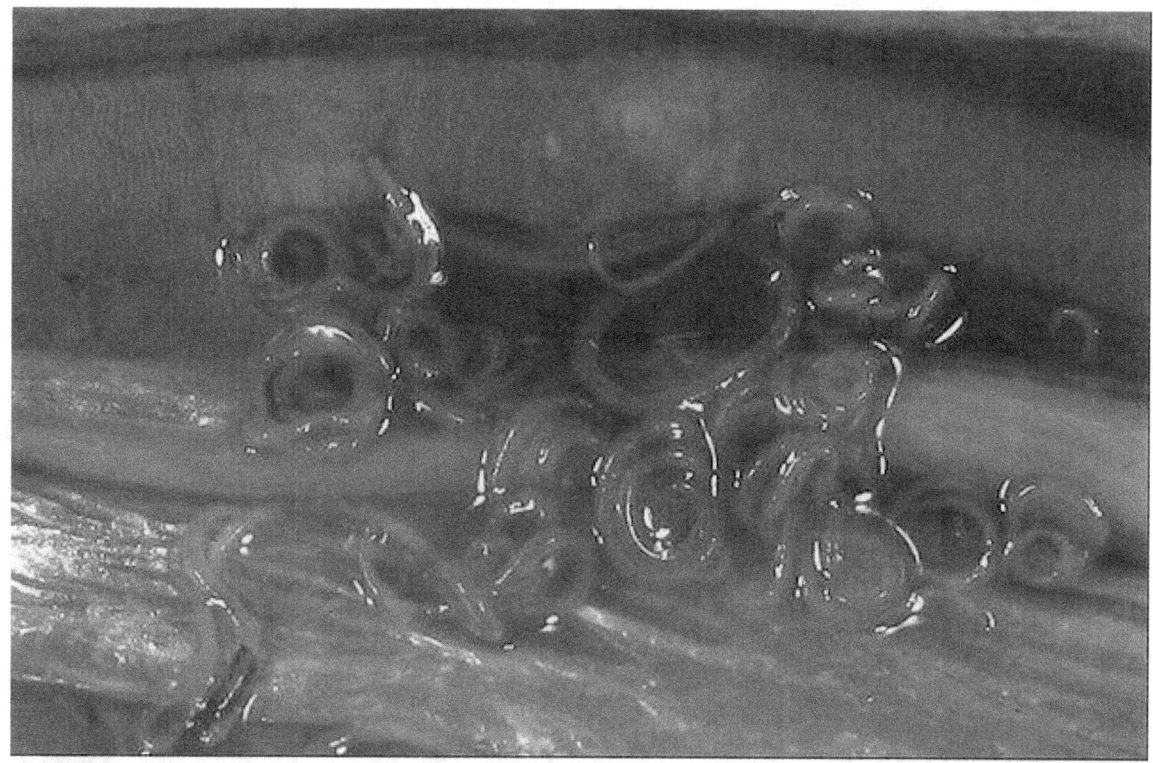

Unidentified Anisakidae (Ascaridina: Ascaridoidea)

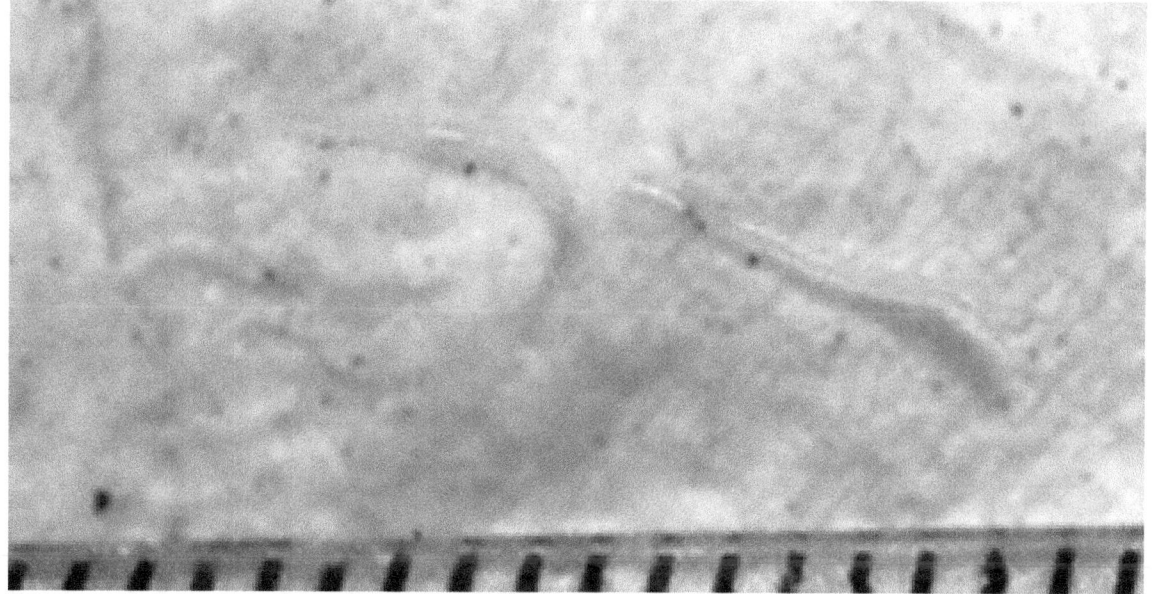

Oxyuridae *Threadworm*

Phylum Nematoda

- Basal order Monhysterida
- Class Dorylaimea
- Class Enoplea

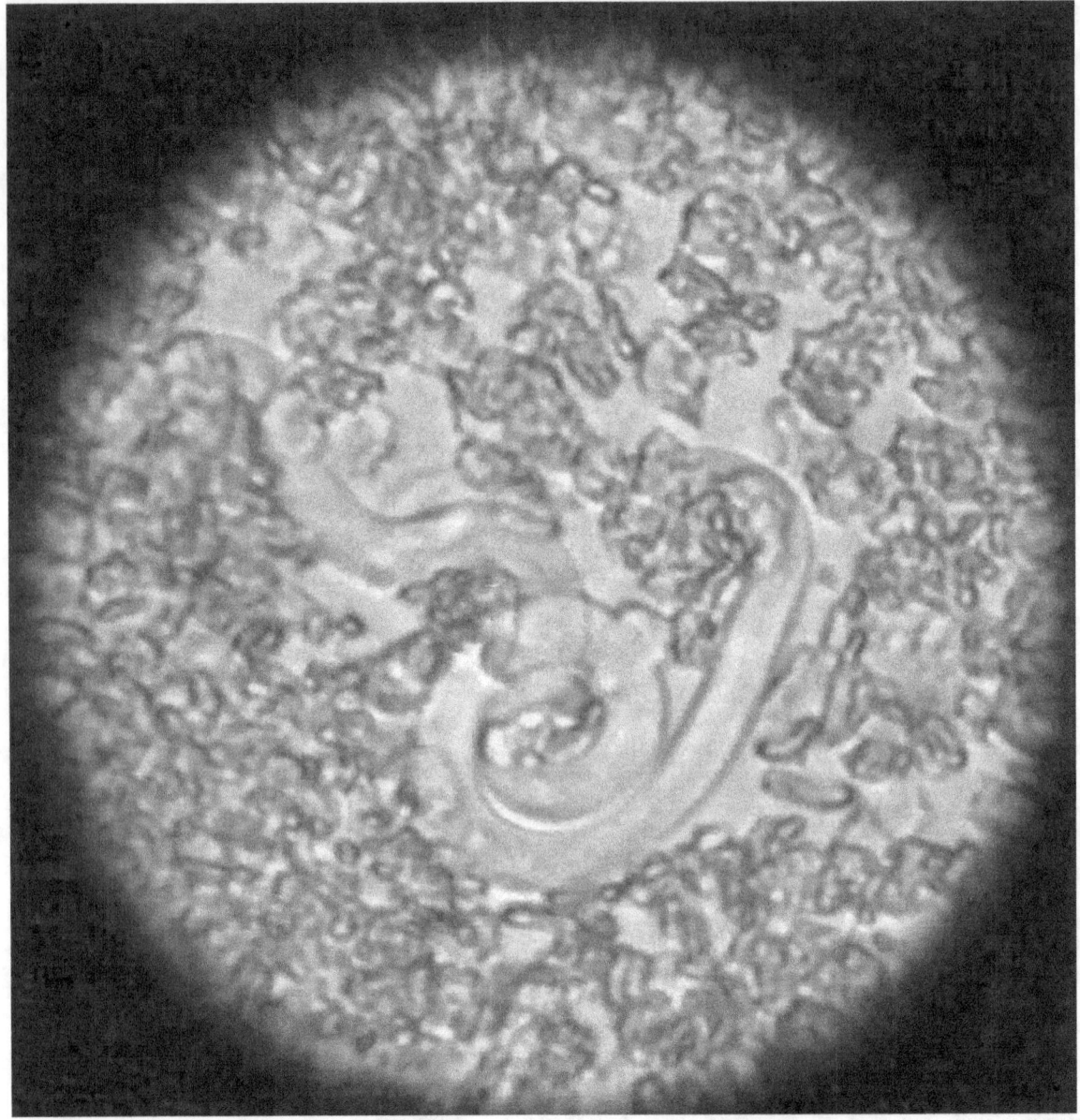

Spiruridae Dirofilaria immitis

- Class Secernentea
  - Subclass Diplogasteria (disputed)
  - Subclass Rhabditia (paraphyletic?)
  - Subclass Spiruria
  - Subclass Tylenchia (disputed)
- "Chromadorea" assemblage

Later work has suggested the presence of 12 clades.*[27] The Secernentea—a group that includes virtually all major animal and plant 'nematode' parasites—apparently arose from within the Adenophorea.

A major effort to improve the systematics of this phylum is in progress and being organised by the 959 Nematode Genomes.*[28]

A complete checklist of the World's nematode species can be found in the World Species Index:Nematoda.*[29]

An analysis of the mitochondrial DNA suggests that the following groupings are valid*[30]

- subclass Dorylaimia

- orders Rhabditida, Trichinellida and Mermithida

- suborder Rhabditina

- infraorders Spiruromorpha and Oxyuridomorpha

The Ascaridomorpha, Rhabditomorpha and Diplogasteromorpha appear to be related.

The suborders Spirurina and Tylenchina and the infraorders Rhabditomorpha, Panagrolaimomorpha and Tylenchomorpha are paraphytic.

The monophyly of the Ascaridomorph is uncertain.

28.2 Anatomy

Nematodes are slender worms: typically approximately 5 to 100 μm thick, and at least 0.1 mm (0.0039 in) but less than 2.5mm long.*[31] The smallest nematodes are microscopic, while free-living species can reach as much as 5 cm (2.0 in), and some parasitic species are larger still, reaching over a meter in length.*[32]*:271 The body is often ornamented with ridges, rings, bristles, or other distinctive structures.*[33]

The head of a nematode is relatively distinct. Whereas the rest of the body is bilaterally symmetrical, the head is radially symmetrical, with sensory bristles and, in many cases, solid 'head-shields' radiating outwards around the mouth. The mouth has either three or six lips, which often bear a series of teeth on their inner edges. An adhesive 'caudal gland' is often found at the tip of the tail.*[34]

The epidermis is either a syncytium or a single layer of cells, and is covered by a thick collagenous cuticle. The cuticle is often of complex structure, and may have two or three distinct layers. Underneath the epidermis lies a layer of longitudinal muscle cells. The relatively rigid cuticle works with the muscles to create a hydroskeleton as nematodes lack circumferential muscles. Projections run from the inner surface of muscle cells towards the nerve cords; this is a unique arrangement in the animal kingdom, in which nerve cells normally extend fibres into the muscles rather than *vice versa.*[34]

28.2.1 Digestive system

The oral cavity is lined with cuticle, which is often strengthened with ridges or other structures, and, especially in carnivorous species, may bear a number of teeth. The mouth often includes a sharp stylet, which the animal can thrust into its prey. In some species, the stylet is hollow, and can be used to suck liquids from plants or animals.*[34]

The oral cavity opens into a muscular, sucking pharynx, also lined with cuticle. Digestive glands are found in this region of the gut, producing enzymes that start to break down the food. In stylet-bearing species, these may even be injected into the prey.*[34]

There is no stomach, with the pharynx connecting directly to a muscleless intestine that forms the main length of the gut. This produces further enzymes, and also absorbs nutrients through its single cell thick lining. The last portion of the intestine is lined by cuticle, forming a rectum, which expels waste through the anus just below and in front of the tip of the tail. Movement of food through the digestive system is the result of body movements of the worm. The intestine has valves or sphincters at either end to help control the movement of food through the body.*[34]

28.2.2   Excretory system

Nitrogenous waste is excreted in the form of ammonia through the body wall, and is not associated with any specific organs. However, the structures for excreting salt to maintain osmoregulation are typically more complex.*[34]

In many marine nematodes, one or two unicellular 'renette glands' excrete salt through a pore on the underside of the animal, close to the pharynx. In most other nematodes, these specialised cells have been replaced by an organ consisting of two parallel ducts connected by a single transverse duct. This transverse duct opens into a common canal that runs to the excretory pore.*[34]

28.2.3   Nervous system

See also: Muscle arms

Four peripheral nerves run the length of the body on the dorsal, ventral, and lateral surfaces. Each nerve lies within a cord of connective tissue lying beneath the cuticle and between the muscle cells. The ventral nerve is the largest, and has a double structure forward of the excretory pore. The dorsal nerve is responsible for motor control, while the lateral nerves are sensory, and the ventral combines both functions.*[34]

The nervous system is also the only place in the nematode body that contains cilia, which are all non-motile and with a sensory function.*[35]*[36]

At the anterior end of the animal, the nerves branch from a dense, circular nerve ring surrounding the pharynx, and serving as the brain. Smaller nerves run forward from the ring to supply the sensory organs of the head.*[34]

The bodies of nematodes are covered in numerous sensory bristles and papillae that together provide a sense of touch. Behind the sensory bristles on the head lie two small pits, or 'amphids'. These are well supplied with nerve cells, and are probably chemoreception organs. A few aquatic nematodes possess what appear to be pigmented eye-spots, but is unclear whether or not these are actually sensory in nature.*[34]

28.3   Reproduction

Most nematode species are dioecious, with separate male and female individuals. Both sexes possess one or two tubular gonads. In males, the sperm are produced at the end of the gonad, and migrate along its length as they mature. The testes each open into a relatively wide sperm duct and then into a glandular and muscular ejaculatory duct associated with the cloaca. In females, the ovaries each open into an oviduct and then a glandular uterus. The uteri both open into a common vagina, usually located in the middle of the ventral surface.*[34]

Reproduction is usually sexual. Males are usually smaller than females (often much smaller) and often have a characteristically bent tail for holding the female. During copulation, one or more chitinized spicules move out of the cloaca and are inserted into the genital pore of the female. Amoeboid sperm crawl along the spicule into the female worm. Nematode sperm is thought to be the only eukaryotic cell without the globular protein G-actin.

Eggs may be embryonated or unembryonated when passed by the female, meaning their fertilized eggs may not yet be developed. A few species are known to be ovoviviparous. The eggs are protected by an outer shell, secreted by the uterus. In free-living roundworms, the eggs hatch into larvae, which appear essentially identical to the adults, except for an underdeveloped reproductive system; in parasitic roundworms, the life cycle is often much more complicated.*[34]

Nematodes as a whole possess a wide range of modes of reproduction.*[38] Some nematodes, such as *Heterorhabditis* spp., undergo a process called *endotokia matricida*: intrauterine birth causing maternal death.*[39] Some nematodes are hermaphroditic, and keep their self-fertilized eggs inside the uterus until they hatch. The juvenile nematodes will then ingest the parent nematode. This process is significantly promoted in environments with a low food supply.*[39]

The nematode model species *Caenorhabditis elegans* and *C. briggsae* exhibit androdioecy, which is very rare among animals. The single genus *Meloidogyne* (root-knot nematodes) exhibit a range of reproductive modes, including sexual reproduction, facultative sexuality (in which most, but not all, generations reproduce asexually), and both meiotic and mitotic parthenogenesis.

The genus *Mesorhabditis* exhibits an unusual form of parthenogenesis, in which sperm-producing males copulate with females, but the sperm do not fuse with the ovum. Contact with the sperm is essential for the ovum to begin dividing, but because there is no fusion of the cells, the male contributes no genetic material to the offspring, which are essentially clones of the female.[*][34]

28.4 Free-living species

In free-living species, development usually consists of four molts of the cuticle during growth. Different species feed on materials as varied as algae, fungi, small animals, fecal matter, dead organisms and living tissues. Free-living marine nematodes are important and abundant members of the meiobenthos. They play an important role in the decomposition process, aid in recycling of nutrients in marine environments, and are sensitive to changes in the environment caused by pollution. One roundworm of note, *Caenorhabditis elegans*, lives in the soil and has found much use as a model organism. *C. elegans* has had its entire genome sequenced, as well as the developmental fate of every cell determined, and every neuron mapped.

28.5 Parasitic species

Nematodes commonly parasitic on humans include ascarids (*Ascaris*), filarias, hookworms, pinworms (*Enterobius*) and whipworms (*Trichuris trichiura*). The species *Trichinella spiralis*, commonly known as the 'trichina worm', occurs in rats, pigs, and humans, and is responsible for the disease trichinosis. *Baylisascaris* usually infests wild animals, but can be deadly to humans, as well. *Dirofilaria immitis* are known for causing heartworm disease by inhabiting the hearts, arteries, and lungs of dogs and some cats. *Haemonchus contortus* is one of the most abundant infectious agents in sheep around the world, causing great economic damage to sheep. In contrast, entomopathogenic nematodes parasitize insects and are mostly considered beneficial by humans, but some attack beneficial insects.

One form of nematode is entirely dependent upon fig wasps, which are the sole source of fig fertilization. They prey upon the wasps, riding them from the ripe fig of the wasp's birth to the fig flower of its death, where they kill the wasp, and their offspring await the birth of the next generation of wasps as the fig ripens.

A newly discovered parasitic tetradonematid nematode, *Myrmeconema neotropicum*, apparently induces fruit mimicry in the tropical ant *Cephalotes atratus*. Infected ants develop bright red gasters (abdomens), tend to be more sluggish, and walk with their gasters in a conspicuous elevated position. It is likely that these changes cause frugivorous birds to confuse the infected ants for berries, and eat them. Parasite eggs passed in the bird's feces are subsequently collected by foraging *Cephalotes atratus* and are fed to their larvae, thus completing the life cycle of *M. neotropicum*.[*][40]

Plant-parasitic nematodes include several groups causing severe crop losses. The most common genera are *Aphelenchoides* (foliar nematodes), *Ditylenchus*, *Globodera* (potato cyst nematodes), *Heterodera* (soybean cyst nematodes), *Longidorus*, *Meloidogyne* (root-knot nematodes), *Nacobbus*, *Pratylenchus* (lesion nematodes), *Trichodorus* and *Xiphinema* (dagger nematodes). Several phytoparasitic nematode species cause histological damages to roots, including the formation of visible galls (e.g. by root-knot nematodes), which are useful characters for their diagnostic in the field. Some nematode species transmit plant viruses through their feeding activity on roots. One of them is *Xiphinema index*, vector of grapevine fanleaf virus, an important disease of grapes, another one is *Xiphinema diversicaudatum*, vector of arabis mosaic virus.

Other nematodes attack bark and forest trees. The most important representative of this group is *Bursaphelenchus xylophilus*, the pine wood nematode, present in Asia and America and recently discovered in Europe.

28.5.1 Agriculture and horticulture

Depending on the species, a nematode may be beneficial or detrimental to plant health. From agricultural and horticulture perspectives, the two categories of nematodes are the predatory ones, which will kill garden pests like cutworms and corn earworm moths, and the pest nematodes, like the root-knot nematode, which attack plants, and those that act as vectors spreading plant viruses between crop plants.*[41] Predatory nematodes can be bred by soaking a specific recipe of leaves and other detritus in water, in a dark, cool place, and can even be purchased as an organic form of pest control.

Rotations of plants with nematode-resistant species or varieties is one means of managing parasitic nematode infestations. For example, marigolds, grown over one or more seasons (the effect is cumulative), can be used to control nematodes.*[42] Another is treatment with natural antagonists such as the fungus *Gliocladium roseum*. Chitosan, a natural biocontrol, elicits plant defense responses to destroy parasitic cyst nematodes on roots of soybean, corn, sugar beet, potato and tomato crops without harming beneficial nematodes in the soil.*[43] Soil steaming is an efficient method to kill nematodes before planting a crop, but indiscriminately eliminates both harmful and beneficial soil fauna.

The Golden Nematode (Globodera rostochiensis) is a particularly harmful variety of nematode pest that has resulted in quarantines and crop failures worldwide. CSIRO has found*[44] a 13- to 14-fold reduction of nematode population densities in plots having Indian mustard (*Brassica juncea*) green manure or seed meal in the soil.

Hundreds of *Caenorhabditis elegans* were featured in a research project on NASA's STS-107 space mission, and were known to have survived the Space Shuttle Columbia disaster.*[45]

28.6 Epidemiology

A number of intestinal nematodes cause diseases affecting human beings, including ascariasis, trichuriasis and hookworm disease. Filarial nematodes cause filariasis.

28.7 Soil ecosystems

90 percent of nematodes reside in the top 15 cm of soil. Nematodes do not decompose organic matter, but, instead, are parasitic and free-living organisms that feed on living material. Nematodes can effectively regulate bacterial population and community composition - they may eat up to 5,000 bacteria per minute. Also, Nematodes can play an important role in the nitrogen cycle by way of nitrogen mineralization.*[31]

28.8 Trivia

Hundreds of nematode worms (C. elegans), featuring in a research project on mission STS-107, survived the Space Shuttle Columbia Disaster [1].

In the SpongeBob SquarePants episode, "Home Sweet Pineapple", his house is consumed by a swarm of nematodes. They appeared again in the episode, "Best Day Ever".

On the BBC2 quiz show QI, when Clive Anderson was asked, "What lives in the Dead Sea?", he answered, "I'm tempted to say nematode worms because they live everywhere."

28.9 See also

- Biological pest control

- Capillaria

- *Caenorhabditis elegans*: An important model organism often used to study cellular differentiation, sometimes simply referred to as "worm" by scientists

- List of organic gardening and farming topics

- List of parasites of humans

- Toxocariasis: A helminth infection of humans caused by the dog or cat roundworm, *Toxocara canis* or *Toxocara cati*

28.10 Notes

[1] Note that words as nematologist, nematosis, nematocide, etc. are based on *nema, nematos* and not on "nematode".

28.11 References

[1] "Nematode Fossils". *Nematode Fossils [Nematoda]*. N.p., n.d. Web. 21 Apr. 2013.

[2] Hodda, M (2011). "Phylum Nematoda Cobb, 1932. In: Zhang, Z.-Q. (Ed.) Animal biodiversity: An outline of higher-level classification and survey of taxonomic richness". *Zootaxa* **3148**: 63–95.

[3] Zhang, Z (2013). "Animal biodiversity: An update of classification and diversity in 2013. In: Zhang, Z.-Q. (Ed.) Animal Biodiversity: An Outline of Higher-level Classification and Survey of Taxonomic Richness (Addenda 2013)". *Zootaxa* **3703** (1): 5–11. doi:10.11646/zootaxa.3703.1.3.

[4] Lambshead PJD (1993). "Recent developments in marine benthic biodiversity research". *Oceanis* **19** (6): 5–24.

[5] Borgonie G, García-Moyano A, Litthauer D, Bert W, Bester A, van Heerden E, Möller C, Erasmus M, Onstott TC (June 2011). "Nematoda from the terrestrial deep subsurface of South Africa". *Nature* **474** (7349): 79–82. doi:10.1038/nature09974. PMID 21637257.

[6] Danovaro R, Gambi C, Dell'Anno A, Corinaldesi C, Fraschetti S, Vanreusel A, Vincx M, Gooday AJ (January 2008). "Exponential decline of deep-sea ecosystem functioning linked to benthic biodiversity loss". *Curr. Biol.* **18** (1): 1–8. doi:10.1016/j.cub.2007.11.056. PMID 18164201. Lay summary – *EurekAlert!*.

[7] Platt HM (1994). "foreword". In Lorenzen S, Lorenzen SA. *The phylogenetic systematics of freeliving nematodes*. London: The Ray Society. ISBN 0-903874-22-9.

[8] Lemonick MD (2011-06-08). "Could 'worms from Hell' mean there's life in space?". *Time*. ISSN 0040-781X. Retrieved 2011-06-08.

[9] Bhanoo SN (2011-06-01). "Nematode found in mine is first subsurface multicellular organism". *The New York Times*. ISSN 0362-4331. Retrieved 2011-06-13.

[10] "Gold mine". *Nature* **474** (7349): 6. June 2011. doi:10.1038/474006b.

[11] Drake N (2011-06-01). "Subterranean worms from hell: Nature News". *Nature News*. Retrieved 2011-06-13.

[12] Borgonie G, García-Moyano A, Litthauer D, Bert W, Bester A, van Heerden E, Möller C, Erasmus M, Onstott TC (2011-06-02). "Nematoda from the terrestrial deep subsurface of South Africa". *Nature* **474** (7349): 79–82. doi:10.1038/nature09974. ISSN 0028-0836. PMID 21637257.

[13] Hsueh YP, Leighton DHW, Sternberg PW. (2014). Nematode Communication. In: Witzany G (ed). Biocommunication of Animals. Springer, 383-407. ISBN 978-94-007-7413-1.

[14] Pramer C (1964). "Nematode-trapping fungi". *Science* **144** (3617): 382–388. doi:10.1126/science.144.3617.382. PMID 14169325.

[15] Hauser JT (December 1985). "Nematode-trapping fungi" (PDF). *Carnivorous Plant Newsletter* **14** (1): 8–11.

[16] Ahrén D, Ursing BM, Tunlid A (1998). "Phylogeny of nematode-trapping fungi based on 18S rDNA sequences". *FEMS Microbiology Letters* **158** (2): 179–184. doi:10.1016/s0378-1097(97)00519-3. PMID 9465391.

[17] Cobb, Nathan (1914). "Nematodes and their relationships". *Yearbook United States Department of Agriculture*. United States Department of Agriculture. pp. 457–90. Quote on p. 472.

[18] Chitwood BG (1957). "The English word "Nema" Revised". *Systematic Zoology in Nematology Newsletter* **4** (45): 1619. doi:10.2307/sysbio/6.4.184.

[19] Siddiqi MR (2000). *Tylenchida: parasites of plants and insects*. Wallingford, Oxon, UK: CABI Pub. ISBN 0-85199-202-1.

[20] "ITIS report: Nematoda". Itis.gov. Retrieved 2012-06-12.

[21] "Bilateria". Tree of Life Web Project (ToL). January 1, 2002. Retrieved 2008-11-02.

[22] Chitwood BG, Chitwood MB (1933). "The characters of a protonematode". *J Parasitol* **20**: 130.

[23] Chitwood BG (1937). "A revised classification of the *Nematoda*". *Papers on helminthology, 30 year jubileum K.J. Skrjabin*. Moscow: All-Union Lenin Academy of Agricultural Sciences. pp. 67–79.

[24] Chitwood BG (1958). "The designation of official names for higher taxa of invertebrates". *Bull Zool Nomencl* **15**: 860–95.

[25] Blaxter ML, De Ley P, Garey JR, Liu LX, Scheldeman P, Vierstraete A, Vanfleteren JR, Mackey LY, Dorris M, Frisse LM, Vida JT, Thomas WK (March 1998). "A molecular evolutionary framework for the phylum Nematoda". *Nature* **392** (6671): 71–5. doi:10.1038/32160. PMID 9510248.

[26] "Nematoda". Tree of Life Web Project (ToL). 2002-01-01. Retrieved 2008-11-02.

[27] Holterman M, van der Wurff A, van den Elsen S, van Megen H, Bongers T, Holovachov O, Bakker J, Helder J (2006). "Phylum-wide analysis of SSU rDNA reveals deep phylogenetic relationships among nematodes and accelerated evolution toward crown Clades". *Mol Biol Evol* **23** (9): 1792–1800. doi:10.1093/molbev/msl044. PMID 16790472.

[28] "959 Nematode Genomes – NematodeGenomes". Nematodes.org. 2011-11-11. Retrieved 2012-06-12.

[29] *World Species Index:Nematoda*. 2012.

[30] Liu, GH; Shao, R; Li, JY; Zhou, DH; Li, H; Zhu, XQ (2013). "The complete mitochondrial genomes of three parasitic nematodes of birds: a unique gene order and insights into nematode phylogeny". *BMC Genomics* **14** (1): 414. doi:10.1186/1471-2164-14-414.

[31] Nyle C. Brady & Ray R. Weil (2009). *Elements of the Nature and Properties of Soils (3rd Edition)*. Prentice Hall. ISBN 9780135014332.

[32] Ruppert EE, Fox RS, Barnes RD (2004). *Invertebrate Zoology: A Functional Evolutionary Approach* (7th ed.). Belmont, California: Brooks/Cole. ISBN 978-0-03-025982-1.

[33] Weischer B, Brown DJF (2000). *An Introduction to Nematodes: General Nematology*. Sofia, Bulgaria: Pensoft. pp. 75–76. ISBN 978-954-642-087-9.

[34] Barnes RG (1980). *Invertebrate zoology*. Philadelphia: Sanders College. ISBN 0-03-056747-5.

[35] The sensory cilia of Caenorhabditis elegans

[36] Hearing in Drosophila Requires TilB, a Conserved Protein Associated With Ciliary Motility

[37] Lalošević, V.; Lalošević, D.; Capo, I.; Simin, V.; Galfi, A.; Traversa, D. (2013). "High infection rate of zoonotic *Eucoleus aerophilus* infection in foxes from Serbia." *Parasite* **20**: 3. doi:10.1051/parasite/2012003. PMC 3718516. PMID 23340229.

[38] Bell G (1982). *The masterpiece of nature: the evolution and genetics of sexuality*. Berkeley: University of California Press. ISBN 0-520-04583-1.

[39] Johnigk S-A, Ehlers R-U (1999). "*Endotokia matricida* in hermaphrodites of *Heterorhabditis* spp. and the effect of the food supply". *Nematology* **1** (7–8): 717–726. doi:10.1163/156854199508748. ISSN 1388-5545.

[40] Yanoviak SP, Kaspari M, Dudley R, Poinar G (April 2008). "Parasite-induced fruit mimicry in a tropical canopy ant". *Am. Nat.* **171** (4): 536–44. doi:10.1086/528968. PMID 18279076.

[41] Purcell, M., M.W. Johnson, L.M. Lebeck, and A.H. Hara. 1992. Biological control of Helicoverpa zea (Lepidoptera: Noctuidae) with Steinernema carpocapsae (Rhabditida: Steinernematidae) in corn used as a trap crop. Environmental Entomology 21:1441-1447.

[42] Riotte L (1975). *Secrets of companion planting for successful gardening*. p. 7.

[43] US application 2008072494, Stoner RJ, Linden JC, "Micronutrient elicitor for treating nematodes in field crops", published 2008-03-27

[44] "CSIRO Publishing – Australasian plant pathology". www.publish.csiro.au. Retrieved 2010-06-14.

[45] "Worms survived Columbia disaster". *Science/Nature*. BBC News. 2003-04-01.

28.12   Further reading

- Atkinson, H.J. (1973). "The respiratory physiology of the marine nematodes *Enoplus brevis* (Bastian) and *E. communis* (Bastian): I. The influence of oxygen tension and body size" (PDF). *J. Exp. Biol.* **59** (1): 255–266.

- BBC News (2003): Worms survived Columbia disaster. Version of 2003-May-01. Retrieved 2008-Nov-04.

- Gubanov, N.M. (1951). "Giant nematoda from the placenta of Cetacea; *Placentonema gigantissima* nov. gen., nov. sp." ."*. Proc. USSR Acad. Sci.* **77** (6): 1123–1125. [in Russian].

- Kaya, Harry K. et al. (1993). "An Overview of Insect-Parasitic and Entomopathogenic Nematodes". In Bedding, R.A. *Nematodes and the Biological Control of Insect Pests.* Csiro Publishing. ISBN 9780643105911.

- Merck Veterinary Manual (MVM) (2006): Giant kidney worm infection in mink and dogs. Retrieved 2007-FEB-10.

- White JG, Southgate E, Thomson JN, Brenner S (August 1976). "The structure of the ventral nerve cord of Caenorhabditis elegans". *Philos. Trans. R. Soc. Lond., B, Biol. Sci.* **275** (938): 327–48. doi:10.1098/rstb.1976.0086. PMID 8806.

- Lee, Donald L, ed. (2010). *The biology of nematodes.* London: Taylor & Francis. ISBN 0415272114. Retrieved 16 December 2014.

- De Ley, P & Blaxter, M 2004, 'A new system for Nematoda: combining morphological characters with molecular trees, and translating clades into ranks and taxa'. in R Cook & DJ Hunt (eds), Nematology Monographs and Perspectives. vol. 2, E.J. Brill, Leiden, pp. 633–653.

28.13   External links

- Harper Adams University College Nematology Research

- Nematodes/roundworms of man

- http://www.ucmp.berkeley.edu/phyla/ecdysozoa/nematoda.html

- European Society of Nematologists

- Nematode.net: Repository of parasitic nematode sequences.

- http://www.nematodes.org/

- NeMys World free-living Marine Nematodes database

- Nematode Virtual Library

- International Federation of Nematology Societies

- Society of Nematologists

- Australasian Association of Nematologists

- Research on nematodes and longevity

- Nematode on BBC

- Nematode worms in an aquarium

- Phylum Nematoda – nematodes on the UF / *IFAS Featured Creatures Web site

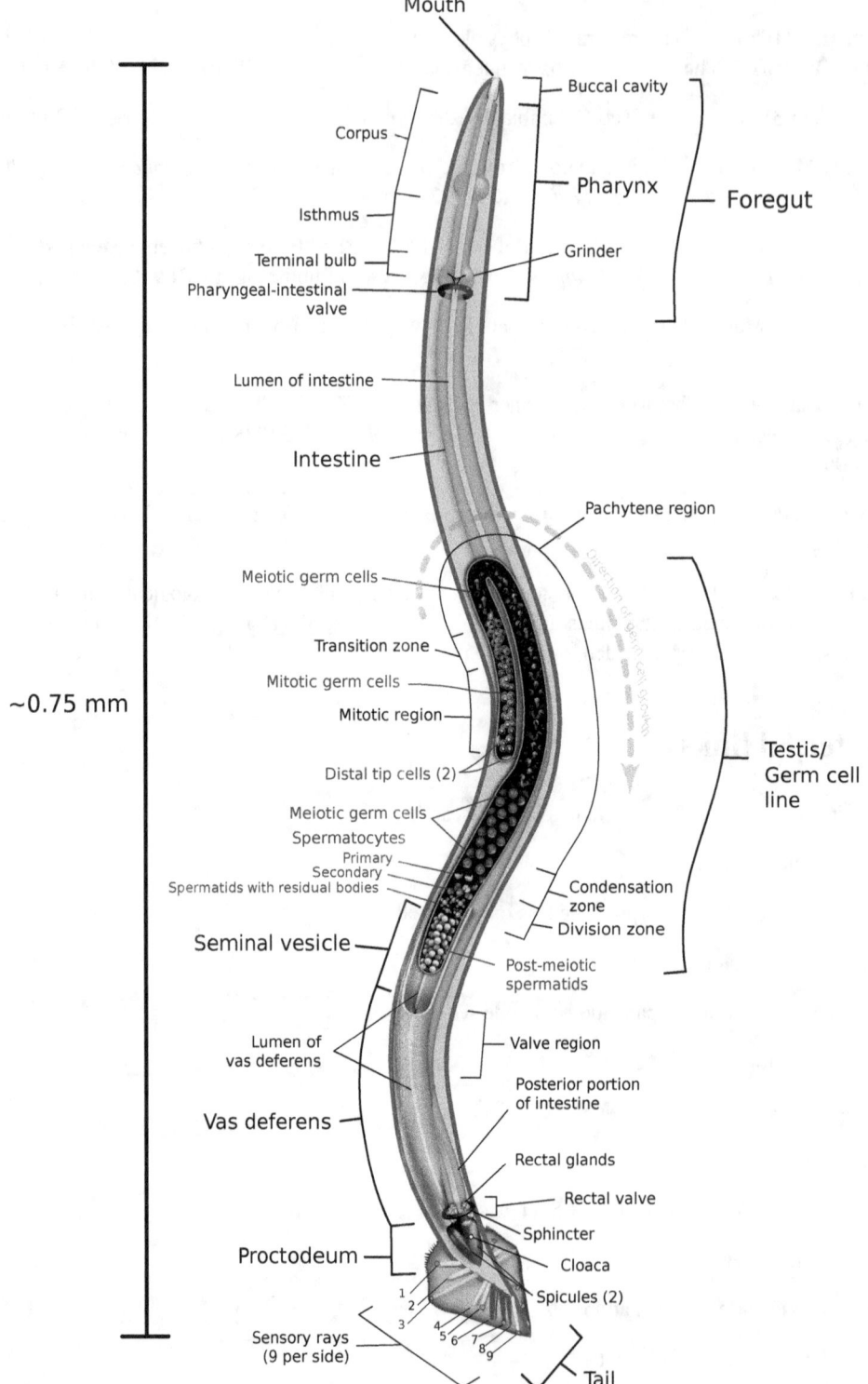

Mouth

Buccal cavity

Corpus

Pharynx

Isthmus

Terminal bulb

Grinder

Foregut

Pharyngeal-intestinal
valve

Lumen of intestine

Intestine

Pachytene region

Meiotic germ cells

Transition zone

Mitotic germ cells

Mitotic region

Distal tip cells (2)

Testis/
Germ cell
line

Meiotic germ cells
Spermatocytes
Primary
Secondary
Spermatids with residual bodies

Condensation
zone

Division zone

Seminal vesicle

Post-meiotic
spermatids

Lumen of
vas deferens

Valve region

Posterior portion
of intestine

Vas deferens

Rectal glands

Rectal valve

Sphincter

Proctodeum

Cloaca

Spicules (2)

1
2
3
4
5 6
7
8 9

Sensory rays
(9 per side)

Tail

~0.75 mm

Internal anatomy of a male C. elegans *nematode*

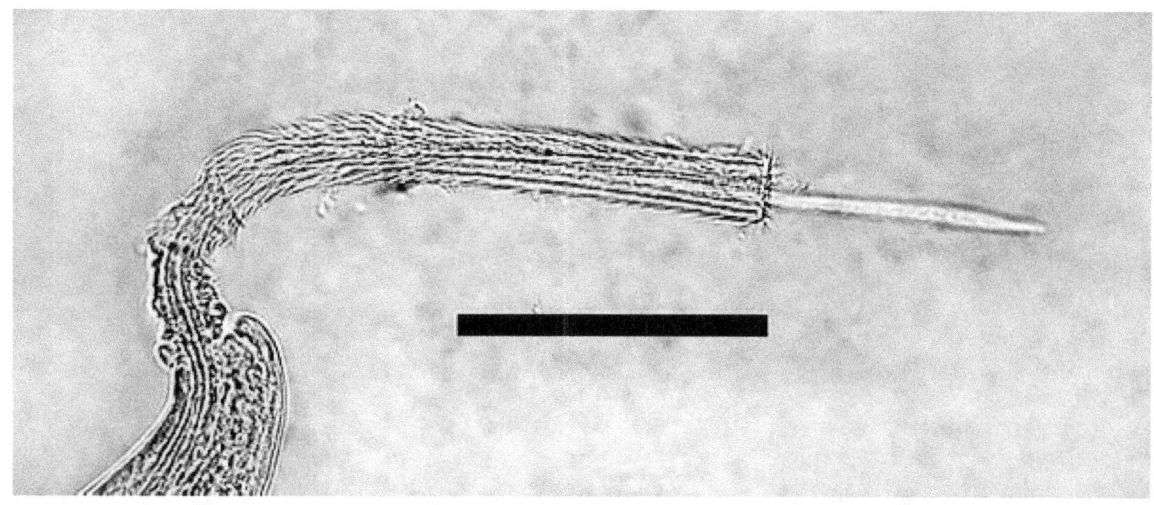

Extremity of a male nematode showing the spicule, used for copulation. Bar = 100 μm [37]

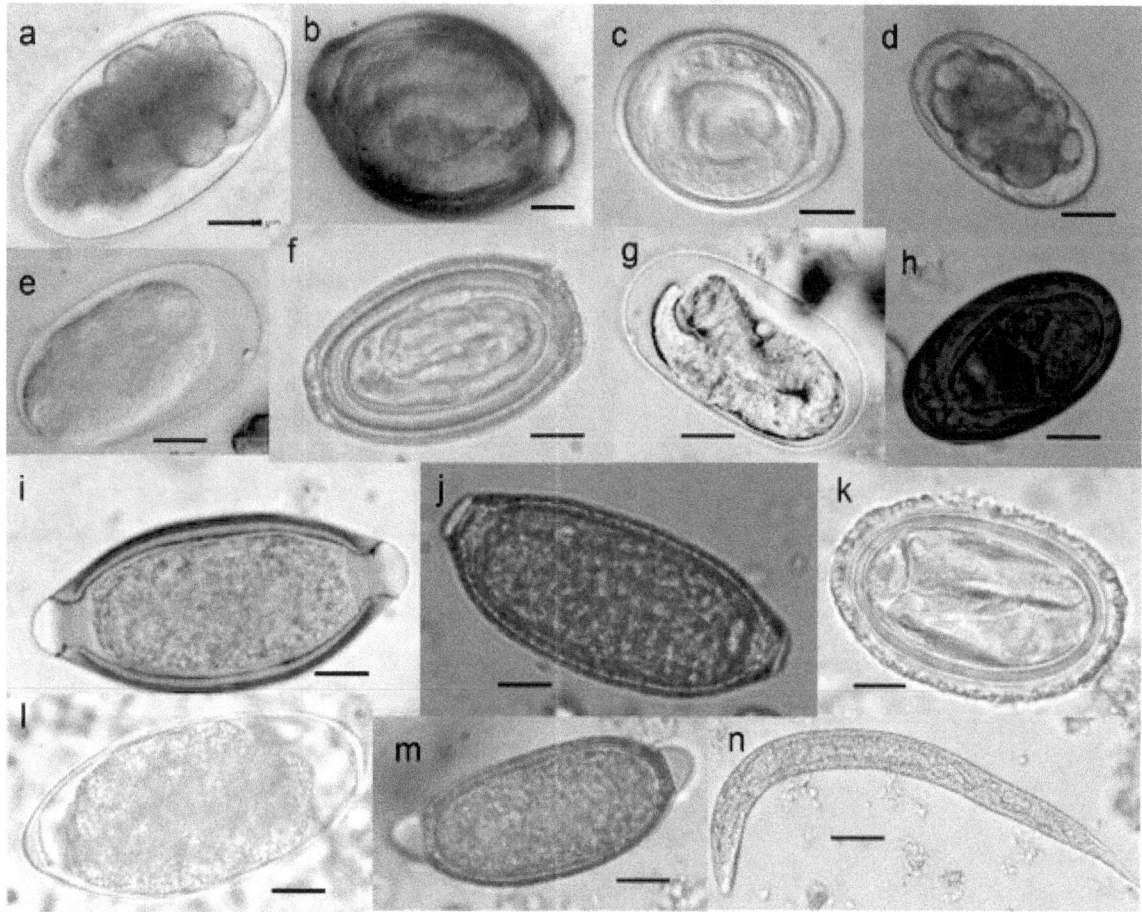

Eggs (mostly nematodes) from stools of wild primates

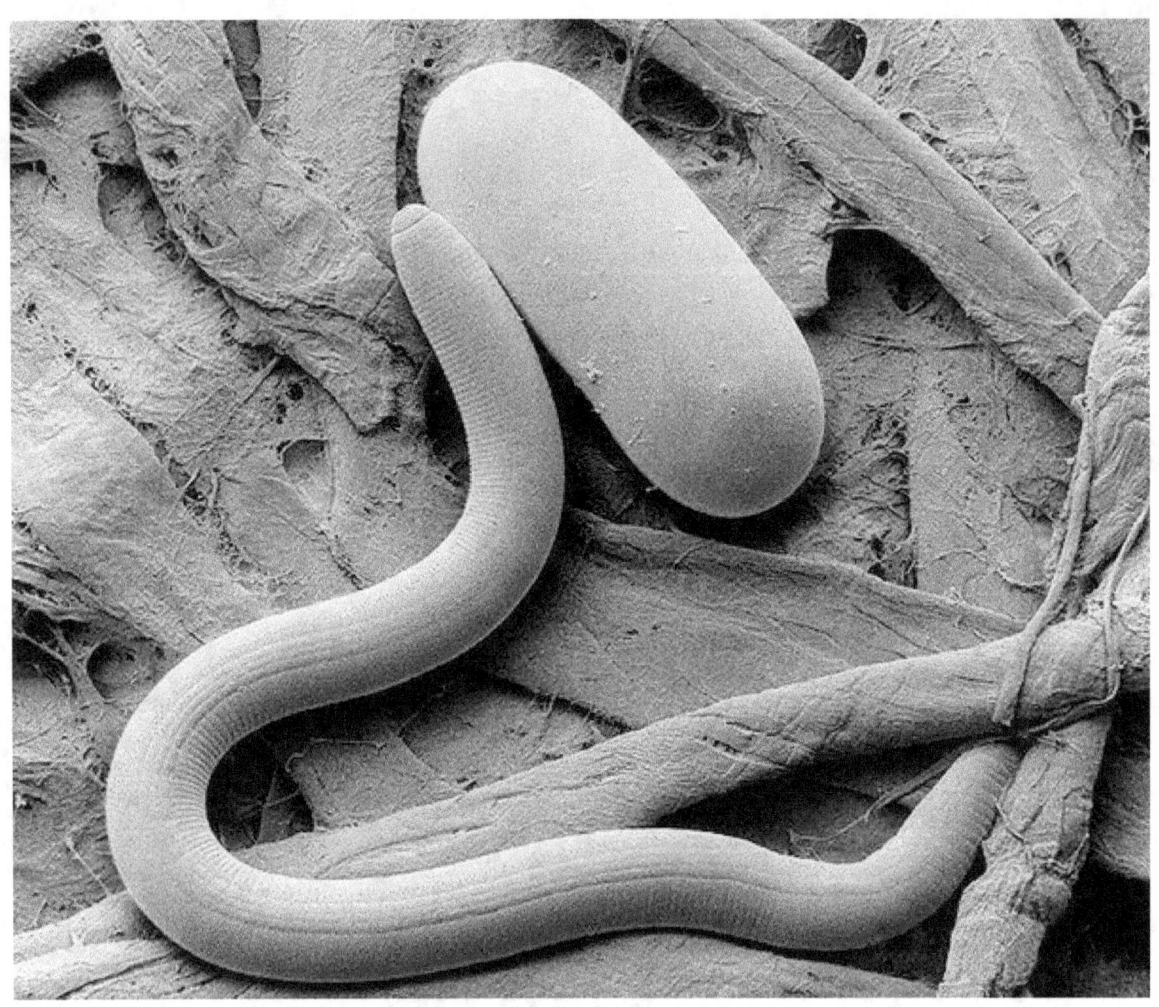

*Colorized electron micrograph of soybean cyst nematode (*Heterodera *sp.) and egg*

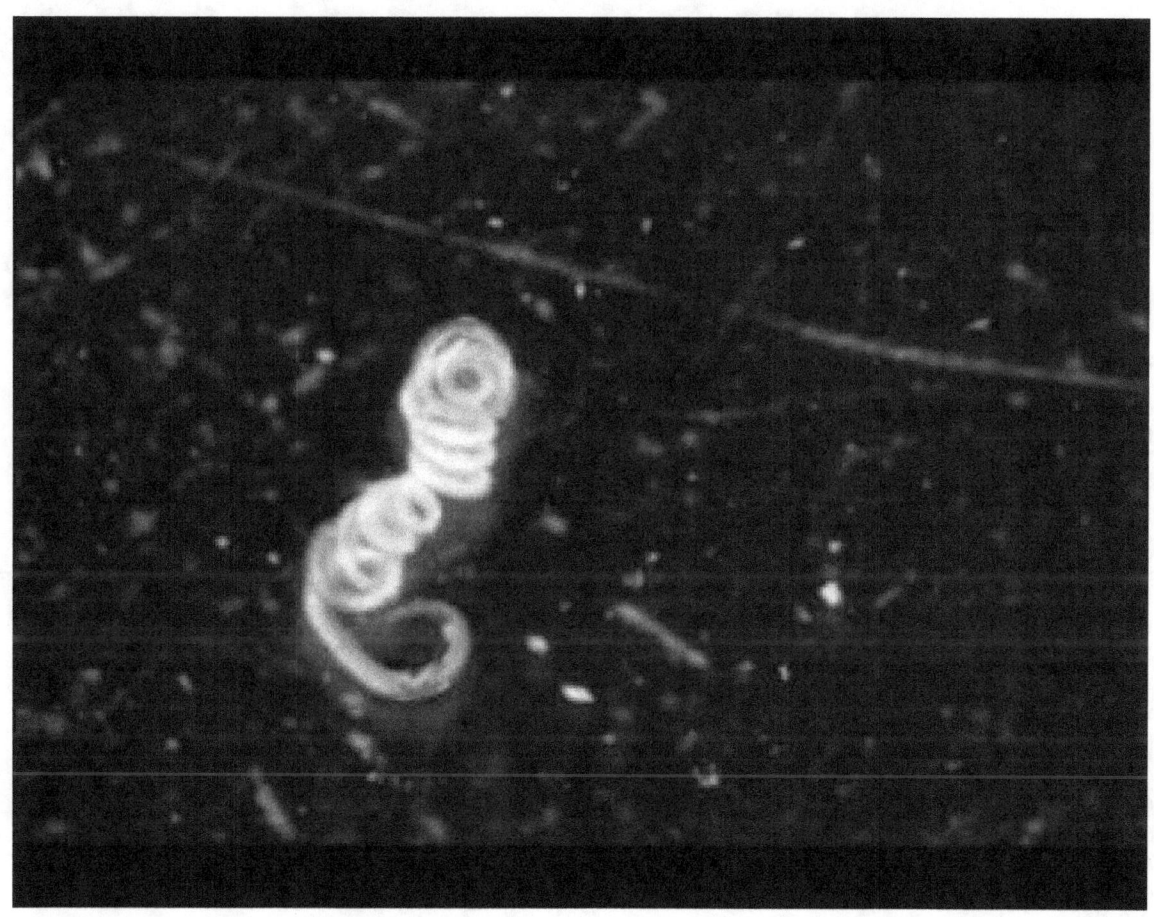

Anthelmintic effect of papain on Heligmosomoides bakeri

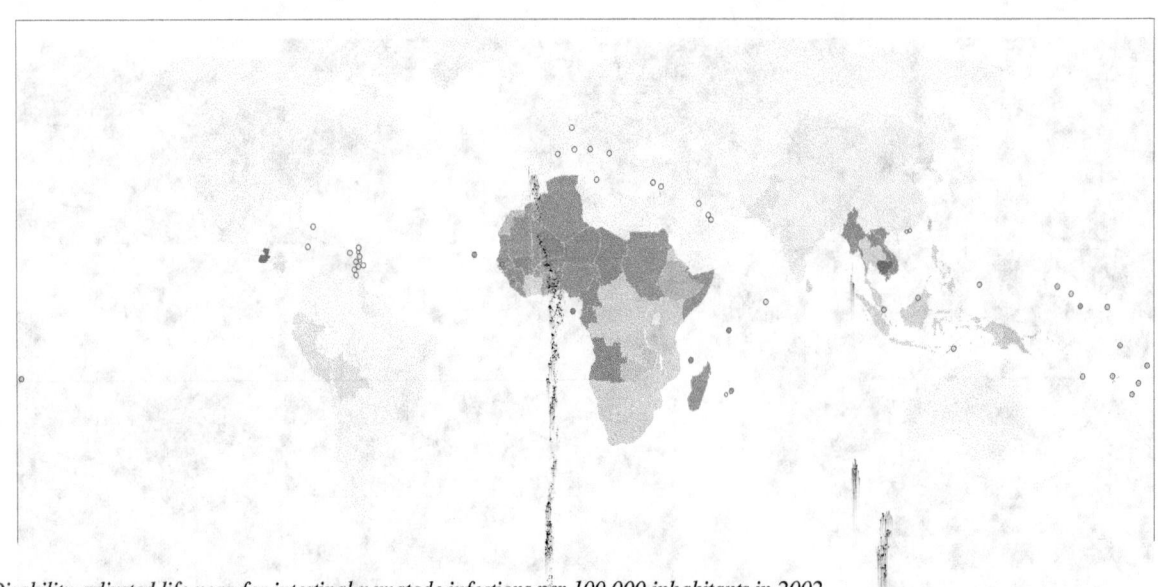

Disability-adjusted life year for intestinal nematode infections per 100,000 inhabitants in 2002.
no data
less than 25
25–50
50–75
75–100
100–120
120–140
140–160
160–180
180–200
200–220
220–240
more than 240

Chapter 29

Night soil

18th-century London nightman's calling card

Night soil is a euphemism for human feces collected at night from cesspools, privies, etc. and sometimes used as a

127

fertilizer. Another definition is "untreated excreta transported without water (e.g. via containers or buckets)".[*][1] Night soil is produced as a result of a sanitation system in areas without community infrastructure such as a sewage treatment facility, or individual septic disposal. In this system of waste management, the human feces are collected in solid form.

29.1 Waste management

29.1.1 Collection

Faeces are excreted into a container or bucket, and are sometimes collected in the container with urine and other waste. The excrement in the pail was often covered with earth/dirt/soil. This may have contributed to the "soil" part of the term "night soil." Often the deposition or excretion occurs within the residence, such as in a shophouse faced with overpopulation. This system is used in isolated rural areas and is important in developing nations or in areas that lack the adequate infrastructure to have running water. The material is collected for temporary storage and is disposed of depending on local custom.

29.1.2 Disposal

Disposal has varied through time. In urban areas, usually slums, a night soil collector will arrive regularly, at varying time periods depending on the supply and demand for night soil collection. Usually this occurs during the night, giving the night soil its name.

In isolated rural areas such as in farms, the household will usually dispose of the night soil themselves, but this practice is generally not referred to as night soil, though the eventual fate of the night soil, and style of handling, is similar.

After arriving at a collection point, usually as a special treatment center within the city, or perhaps an open cesspit, methods of dealing with the waste vary. The waste may go on being shipped to another larger centre to be ultimately taken care of, or be disposed of at that particular juncture.

29.2 Sanitation issues

The use of unprocessed human feces as fertilizer is a risky practice as it may contain disease-causing pathogens. Nevertheless, in developing nations it is widespread. Common parasitic worm infections, such as ascariasis, in these countries are linked to night soil, because their eggs are in feces.

These risks are eliminated by proper composting. "Finished compost should never be 'sterile,' but it should be sanitary. That means the compost should be teeming with beneficial microorganisms that do not pose a threat to human health. Any human disease organisms that may have been in the original organic material should have been eliminated, weakened, or greatly diminished by the time the compost has become mature." [*][2]

Human waste may be attractive as fertilizer because of the high demand for fertilizer and the relative availability of the material to create night soil. In areas where native soil is of poor quality, the local population may weigh the risk of using night soil.

The safe reduction of human waste into compost is possible. Many municipalities create compost from the sewage system biosolids, but then recommend that it only be used on flower beds, not vegetable gardens. Some claims have been made that this is dangerous or inappropriate without the expensive removal of heavy metals.

29.3 Historical examples

29.3.1 Ancient Attica

The use of sewage as fertilizer was common in ancient Attica. The sewage system of ancient Athens collected the sewage of the city in a large reservoir and then channelled it to the Cephissus river valley for use as fertilizer.[*][3]

29.3.2 United Kingdom

A gong farmer was the term used in Tudor England for a person employed to remove human excrement from privies and cesspits. Gong farmers were only allowed to work at night and the waste they collected had to be taken outside the city or town boundaries. They later became known as "night soil men" or "nightmen". In the Manchester area they were also known as the Midnight Mechanic.

29.3.3 India

People responsible for the disposal of night soil are considered untouchables in India. The practice of untouchability was banned by law when India gained independence, but the tradition widely persists as the law is difficult to enforce. This "manual scavenging" is now illegal in all Indian states.

The Indian government's Union Ministry for Social Justice and Empowerment stated in 2003 that 676,000 people were employed in the manual collection of human waste in India. Social organizations have estimated that up to 1.3 million Indians collect such waste. Further, workers in the collection of human waste were confined to marriage amongst themselves, thereby leading to a waste-collecting caste, which passes its profession on from generation to generation.

Employment of Manual Scavengers and Creation of Dry Latrines (Prohibition) Act 1993 has made manual scavenging illegal.

29.3.4 Japan

See also: Toilets in Japan

The reuse of feces as fertilizer was common in Japan. Waste products of rich people were sold at higher prices because their diet was better; therefore, more nutrients remained in their waste. Various historic documents[*][4] dating from the 9th century detail the disposal procedures for toilet waste.

Selling human waste products as fertilizers became much less common after World War II, both for sanitary reasons and because of the proliferation of chemical fertilizers, and less than 1% is used for night soil fertilization. The presence of the United States occupying force, by whom the use of human waste as fertilizer was seen as unhygienic and suspect, was also a contributing factor: "the Occupationaires condemned the practice, and tried to prevent their compatriots from eating vegetables and fruit from the local markets".[*][5]

Modern Japan still has areas with ongoing night soil collection and disposal. The Japanese name for the 'outhouse within the house' style toilet, where night soil is collected for disposal, is Kumitori Benjo (汲み取り便所). The proper disposal or recycling of sewage remains an important research area that is highly political.

29.3.5 China, Singapore, and Hong Kong

The term is known, or even infamous, among the generations that were born in parts of China or Chinatowns (depending on the development of the infrastructure) before 1960. Post-World War II Chinatown, Singapore, before the independence of Singapore, utilized night-soil collection as a primary means of waste disposal, especially as much of the infrastructure was damaged and took a long time to rebuild following the Battle of Singapore and subsequent Japanese Occupation of Singapore. Following the development of the economy and the standard of living after independence, the night soil system in Singapore is now merely a curious anecdote from the time of colonial rule when new systems developed.

The collection method is generally very manual and heavily relies on close human contact with the waste. During the Nationalist era when the Kuomintang ruled mainland China, as well as Chinatown in Singapore, the night soil collector usually arrived with spare and relatively empty honey buckets to exchange for the full honey buckets. The method of transporting the honey buckets from individual households to collection centers was very similar to delivering water supplies by an unskilled laborer, with the exception that the item being transported was not at all potable and it was being delivered *from* the household, rather than *to* the household. The collector would hang full honey buckets onto each end of a pole he carried on his shoulder and then proceeded to carry it through the streets until he reached the collection point.

Hong Kong has a similar euphemism, 倒夜香 *dàoyèxiāng*, which literally means "pour night fragrant".

29.4   See also

- Historical digging

- Humanure

- Jenkem

- Pail closet

- Slopping out

29.5   References

[1] *Sanitation safety planning: manual for safe use and disposal of wastewater, greywater and excreta.* http://www.who.int/water_sanitation_health/publications/ssp-manual/en/: World Health Organization. 2015. ISBN 978 92 4 154924 0.

[2] *Humanure Handbook* (PDF). p. 9.

[3] Durant, Will, *The Life of Greece*, PP. 269

[4] Ebrey, P., Walthall. A., & Palias, J. (2006). Modern east asia: A cultural, social, & political history. Houghton Mifflin Company. Boston & New York. p. 337

[5] "Pictures". Ohio State University. Retrieved 4 December 2010.

29.6   External links

- Necessary and Sufficient - Night soil in colonial America

- Thailand - Pollution from Solid waste and Night Soil

- "Feedback on night Soil". *Mother Earth News*. 1973.

- Arjun Makhijani and Alan Poole. "Vignettes of Third World Agriculture". *Energy and Agriculture in the Third World: A Report to the Energy Policy Project of the Ford Foundation.*

- Dennis T. Avery. "Why Not A Declaration for Sustainable Farming And Forestry?". *Center for Global Food Issues.*

- "Establishment of Small Scale Business Through Night Soil in Kabul".

- Etymology of soil

A night soil man's lamp

Chapter 30

Oligochaeta

This article is about the group of worms. For the plant genus, see Oligochaeta (plant).

Oligochaeta /ˌɒlɪɡəkˈiːtə/ is a class of animals in the phylum Annelida, which is made up of many types of aquatic and terrestrial worms, and including all of the various earthworms. Specifically, the **oligochaetes** /ˈɒlɪɡəkˈiːts/ contain the terrestrial megadrile earthworms (some of which are semiaquatic or fully aquatic), and freshwater or semiterrestrial microdrile forms, including the tubificids, pot worms and ice worms (Enchytraeidae), blackworms (Lumbriculidae) and several interstitial marine worms.

With around 10,000 known species, the Oligochaeta make up about half of the phylum Annelida. These worms usually have few setae (chaetae) or "bristles" on their outer body surfaces, and lack parapodia, unlike polychaeta.

30.1   Common characteristics

Oligochaetes are well-segmented worms and most have a spacious body cavity (coelom) used as a hydroskeleton. They range in length from less than 0.5 mm (0.020 in) up to 2 to 3 metres (6.6 to 9.8 ft) in the 'giant' species such as the giant Gippsland earthworm and the Mekong worm *Amynthas mekongianus* (Cognetti, 1922).[*][1]

The first segment, or prostomium, of oligochaetes is usually a smooth lobe or cone without sensory organs, although it is sometimes extended to form a tentacle. The remaining segments have no appendages, but they do have a small number of bristles, or chaetae. These tend to be longer in aquatic forms than in the burrowing earthworms, and can have a variety of shapes.

Each segment has four bundles of chaetae, with two on the underside, and the others on the sides. The bundles can contain one to 25 chaetae, and include muscles to pull them in and out of the body. This enables the worm to gain a grip on the soil or mud as it burrows into the substrate. When burrowing, the body moves peristaltically, alternately contracting and stretching to push itself forward.

A number of segments in the forward part of the body are modified by the presence of numerous secretory glands. Together, they form the clitellum, which is important in reproduction.[*][2]

30.1.1   Internal anatomy

Most oligochaetes are detritus feeders, although some genera are predaceous, such as *Agriodrilus* and *Phagodrilus*. The digestive tract is essentially a tube running the length of the body, but has a powerful muscular pharynx immediately behind the mouth cavity. In many species, the pharynx simply helps the worm suck in food, but in many aquatic species, it can be turned inside out and placed over food like a suction cup before being pulled back in.

The remainder of the digestive tract may include a crop for storage of food, and a gizzard for grinding it up, although these are not present in all species. The oesophagus includes "calciferous glands" that maintain calcium balance by excreting indigestible calcium carbonate into the gut. A number of yellowish chloragogen cells surround the intestine and the dorsal blood vessel, forming a tissue that functions in a similar fashion to the vertebrate liver. Some of these cells also float freely in the body cavity, where they are referred to as "eleocytes".[2]

Most oligochaetes have no gills or similar structures, and simply breathe through their moist skin. The few exceptions generally have simple, filamentous gills. Excretion is through small ducts known as metanephridia. Terrestrial oligochaetes secrete urea, but the aquatic forms typically secrete ammonia, which dissolves rapidly into the water.[2]

The vascular system consists of two main vessels connected by lateral vessels in each segment. Blood is carried forward in the dorsal vessel (in the upper part of the body) and back through the ventral vessel (underneath), before passing into a sinus surrounding the intestine. Some of the smaller vessels are muscular, effectively forming hearts; from one to five pairs of such hearts is typical. The blood of oligochaetes contains haemoglobin in all but the smallest of species, which have no need of respiratory pigments.[2]

The nervous system consists of two ventral nerve cords, which are usually fused into a single structure, and three or four pairs of smaller nerves per body segment. Only a few aquatic oligochaetes have eyes, and even then they are only simply ocelli. Nonetheless, their skin has several individual photoreceptors, allowing the worm to sense the presence of light, and burrow away from it. Oligochaetes can taste their surroundings using chemoreceptors located in tubercles across their body, and their skin is also supplied with numerous free nerve endings that presumably contribute to their sense of touch.[2]

30.2 Families

- Randiellidae Erséus & Strehlow, 1986

- Naididae / Tubificidae Vejdovsky, 1884 (including Naidinae Ehrenberg, 1831)

- Narapidae Righi, 1983

- Opistocystidae Cernosvitov, 1936

- Dorydrilidae Cook, 1971

- Parvidrilidae Erséus, 1999

- Phreodrilidae Beddard, 1891

- Propappidae Coates, 1986

- Haplotaxidae Michaelsen, 1900

- Tiguassuidae Brinkhurst, 1988

- Lumbriculidae Vejdovsky, 1884

- Enchytraeidae Vejdovsky, 1879

- Moniligastridae Claus, 1880

- Alluroididae Michaelsen, 1900

- Syngenodrilidae Smith & Green, 1919

- Glossoscolecidae Michaelsen, 1900

- Tumakidae Righi, 1995

- Ailoscolecidae Bouché, 1969 (including Komarekionidae Gates, 1974)

- Sparganophilidae Michaelsen, 1918

- Microchaetidae Michaelsen, 1900

- Tritogeniidae Plisko, 2013

- Lumbricidae Claus, 1876 (including Diporodrilinae Bouché, 1970; Eiseniinae Omodeo, 1956; Spermophorodrilinae Omodeo & Rota, 1989; Postandrilinae Qiu & Bouché, 1998; Allolobophorinae Kvavadze, 2000 and Helodrilinae Kvavadze, 2000)

- Kynotidae Brinkhurst & Jamieson, 1971

- Hormogastridae Michaelsen, 1900 (including Vignysinae Bouché, 1970 and Xaninae Diaz Cosin *et al.*, 1989)

- Lutodrilidae McMahan, 1978

- Criodrilidae Vejdovsky, 1884 (including Biwadrilidae Brinkhurst & Jamieson, 1971)

- Almidae Duboscq, 1902

- Ocnerodrilidae Beddard, 1891 (including Malabariinae Gates, 1966)

- Acanthodrilidae Claus, 1880 (including Diplocardiinae Michaelsen, 1900)

- Octochaetidae Michaelsen, 1900 (including Benhamiinae Michaelsen, 1895/7)

- Exxidae Blakemore, 2000

- Megascolecidae Rosa, 1891 (including Pontodrilinae Vejdovsky, 1884; Plutellinae Vejdovsky, 1884 and Argilophilinae Fender & McKey-Fender, 1990)

- Eudrilidae Claus, 1880

30.3   References

[1] Blakemore, Robert J., Csaba Csuzdi, Masamichi T. Ito, Nobuhiro Kaneko, Maurizio G. Paoletti, Sergei E. Spiridonov, Tomoko Uchida & Beverley D. Van Praagh (2007). Megascolex (Promegascolex) mekongianus Cognetti, 1922: its extent, ecology and allocation to Amynthas (Oligochaeta: Megascolecidae). Opuscula Zoologica. 36: 19-30 (Aug. 2007) .

[2] Barnes, Robert D. (1982). *Invertebrate Zoology*. Philadelphia, PA: Holt-Saunders International. pp. 528–547. ISBN 0-03-056747-5.

30.4   Bibliography

- Blakemore, R. J. (2005). Whither Octochaetidae? – its family status reviewed. In: *Advances in Earthworm Taxonomy II*. Eds. A. A. & V. V. Pop. Proceedings IOTM2, Cluj University Press. Romania. Pp. 63–84. http://www.oligochaeta.org/ITOM2/IOTM2.htm.

- Blakemore, R. J. (2006). Revised Key to Earthworm Families (Ch. 9). In: *A Series of Searchable Texts on Earthworm Biodiversity, Ecology and Systematics from Various Regions of the World* – 2nd Edition (2006). Eds.: N. Kaneko & M. T. Ito. COE Soil Ecology Research Group, Yokohama National University, Japan. CD-ROM Publication. Website: http://bio-eco.eis.ynu.ac.jp/eng/database/earthworm/.

- Erséus, C.; Källersjö, M. (2003). "18S rDNA phylogeny of basal groups of Clitellata (Annelida)". *Zoologica Scripta* **33** (2): 187–196. doi:10.1111/j.1463-6409.2004.00146.x.

- Michaelsen, W. (1900). *Das Tierreich 10: Vermes, Oligochaeta*. Friedländer & Sohn, Berlin. Pp. xxix+575, figs. 1-13. Online here: http://mail2web.com/cgi-bin/redir.asp?lid=0&newsite=https://archive.org/details/oligochaeta10mich.

- Plisko, J.D. (2013). A new family Tritogeniidae for the genera *Tritogenia* and *Michalakus*, earlier accredited to the composite Microchaetidae (Annelida: Oligochaeta). *African Invertebrates* **54** (1): 69–92.

- Siddall, M. E., Apakupakul, K, Burreson, E. M., Coates, K. A., Erséus, C, Gelder, S. R., Källersjö, M, & Trapido-Rosenthal, H. (2001). Validating Livanow's Hypothesis: Molecular Data Agree that Leeches, Branchiobdellidans and Acanthobdella peledina form a Monophyletic Group of Oligochaetes. Molecular Phylogenetics and Evolution, 21: 346-351. http://research.amnh.org/~{}siddall/pub/livanow.pdf.

- Stephenson, J. (1930). *The Oligochaeta*. Clarendon Press, Oxford. Pp. 978.

30.5 External links

- Media related to Oligochaeta at Wikimedia Commons

Chapter 31

Olive mill pomace

Olive mill pomace is a by-product from the olive oil mill extraction process. Usually it is used as fuel in a cogeneration system or as organic fertiliser after a composting operation.

Olive mill pomace compost is made by a controlled biologic process that transforms organic waste into a stable humus. Adding composted olive mill pomace as organic fertiliser in olive orchards allows the soil to get nutrients back after each olive crop.

31.1 Two-phase pomace

In crude olive oil production, the traditional system, i.e. pressing, and the three-phase system produce a press cake and a considerable amount of waste water while the two-phase system, which is mainly used in Spain, produces a paste-like waste called "alperujo" or "two-phase pomace" that has a higher water content and is more difficult to treat than traditional solid waste. The water content of the press cake, composed of crude olive cake, pomace and husk, is about 30 percent if it is produced by traditional pressing technology and about 45–50 percent using decanter centrifuges. The press cake still has some oil that is normally recovered in a separate installation. The exhausted olive cake is incinerated or used as a soil conditioner in olive groves.

31.2 External links

- Does the composted olive mill pomace increase the sustainable N use of olive oil cropping?. 2009. 16th Nitrogen workshop. Connecting different scales of Nitrogen use in agriculture. Turin Italia.

Chapter 32

Organopónicos

Produce and flowers from a Cuban organopónico

Organopónicos are a system of urban organic gardens in Cuba. They often consist of low-level concrete walls filled with organic matter and soil, with lines of drip irrigation laid on the surface of the growing media. *Organopónicos* are a labour-intensive form of local agriculture.

Organopónicos first arose as a community response to lack of food security after the collapse of the Soviet Union. They are publicly functioning in terms of ownership, access and management, but heavily subsidized and supported by the Cuban government. Cuba continues to have food rationing, and imports even more food than before.

32.1   Background

During the Cold War, the Cuban economy relied heavily on support from the Soviet Union. In exchange for sugar, Cuba received subsidized oil, chemical fertilizers, pesticides and other farm products. Approximately 50 percent of Cuba's food was imported. Cuba's food production was organized around Soviet-style, large-scale, industrial agricultural collectives.*[1] Before the collapse of the Soviet Union, Cuba used more than 1 million tons of synthetic fertilizers a year and up to 35,000 tons of herbicides and pesticides a year.*[1]

With the collapse of the USSR, Cuba lost its main trading partner and the favorable trade subsidies it received, as well as access to oil, chemical fertilizers, pesticides etc. From 1989 to 1993, the Cuban economy contracted by 35 percent; foreign trade dropped 75 percent.*[1] Without Soviet aid, domestic agriculture production fell by half. This time, called in Cuba the *Special Period*, food scarcities became acute. The average per capita calorie intake fell from 2,900 a day in 1989 to 1,800 calories in 1995. Protein consumption plummeted 40 percent.*[1]

Without food, Cubans had to learn to start growing their own food rather than importing it. This was done through small private farms and thousands of pocket-sized urban market gardens—and, lacking chemicals and fertilizers, food became de facto organic.*[2] Thousands of new urban individual farmers called *parceleros* (for their *parcelas*, or plots) emerged. They formed and developed farmer cooperatives and farmers markets. These urban farmers found the support of the Cuban Ministry of Agriculture (MINAGRI), who provided university experts to train volunteers with organic pesticides and beneficial insects.

Without the fertilizers, hydroponic units from the Soviet Union were no longer usable. The systems were then converted for the use of organic gardening. The original hydroponic units, long cement planting troughs and raised metal containers, were filled with composted sugar waste and *hydroponicos* became *organopónicos*.

The rapid expansion of urban agriculture in the early 1990s included the colonization of vacant land both by community and commercial groups. In Havana, *organopónicos* were created in vacant lots, old parking lots, abandoned building sites and even spaces between roads.

32.2   Current status

More than 35,000 hectares (over 87,000 acres) of land are being used in urban agriculture in Havana alone.*[3] The city of Havana produces enough food for each resident to receive a daily serving of 280 grams (9.88 ounces) of fruits and vegetables. The urban agricultural workforce in Havana has grown from 9,000 in 1999 to 23,000 in 2001 to more than 44,000 in 2006.*[3] However, Cuba still has food rationing for basic staples. Approximately 69% of these rationed basic staples (wheat, vegetable oils, rice, etc.) are imported.*[4] Overall, however, approximately 16% of food is imported from abroad.*[4]

> The grip of the state on Cuban farming has been disastrous. State farms of various kinds hold 75% of Cuba's 6.7m hectares of agricultural land. In 2007 some 45% of this was lying idle, much of it overrun by marabú, a tenacious weed. Cuba is the only country in Latin America where killing a cow is a crime (and eating beef a rare luxury). That has not stopped the cattle herd declining from 7m in 1967 to 4m in 2011.
> —The Economist

The structures of *organopónicos* vary from garden to garden. Some are run by employees of the state; others are run cooperatively by the gardeners themselves. The reliance on the state government cannot be overlooked. The government provides the community farmers with the land and the water. The gardens can buy key materials such as organic composts, seeds and irrigation parts, as well as "biocontrols" such as beneficial insects and plant-based oils that work as pesticides from the government . These biological pest and disease controls are produced in some 200 government centers across the country.*[1]

All garden crops such as beans, tomatoes, bananas, lettuce, okra, eggplant and taro are grown intensively within the city using only organic farming methods since these are the only methods permitted in the urban parts of Havana. No chemicals are used in 68% of Cuban corn, 96% of cassava, 72% of coffee and 40% of bananas. Between 1998 and 2001, chemicals were reduced by 60% in potatoes, 89% in tomatoes, 28% in onion and 43% in tobacco.*[3]

By 1999, some farmers could have black beans, rice, tomato or even a boiled potato to eat; this is impressive by Cuban standards.*[5]

As of 2012 there were plans to privatise farming and dismantle the Organopónicos, as part of broader plans to improve productivity; it is hoped that food rationing could finally end.*[6]

32.3 Applicability beyond Cuba

In Venezuela, the socialist government is trying to introduce urban agriculture to the populace.*[7] In Caracas, the government has launched Organoponico Bolivar I, a pilot program to bring *organopónicos* to Venezuela. Urban agriculture has not been embraced in Caracas as it has in Cuba.*[7] Unlike Cuba, where *organopónicos* arose from the bottom-up out of necessity, the Venezuelan *organopónicos* are clearly a top-down initiative based on Cuba's success. Another problem for urban agriculture in Venezuela is the high amounts of pollution in major Venezuelan urban areas. At the Organoponico Bolivar I, a technician comes every 15 days to take a reading from the small pollution meter in the middle of the garden.*[7]

32.4 See also

- Allotment gardens
- Community Supported Agriculture
- CPA (Agriculture)
- Food security
- Garden sharing
- Guerrilla gardening
- List of community gardens
- Sustainability
- UBPC
- Urban gardening
- Urban horticulture

32.5 References

[1] Mark, Jason (Spring 2007). "Growing it Alone". Earth Island Institute. Retrieved 2010-05-18.

[2] Buncombe, Andrew (August 8, 2006). "The good life in Havana: Cuba's green revolution". *The Independent* (London: Independent Print Limited). Retrieved 2010-05-18.

[3] Knoot, Sinan (January 2009). "The Urban Agriculture of Havana". *Monthly Review* (Monthly Review Foundation) **60**: 44–63. Retrieved 2010-05-18.

[4] "The Paradox of Cuban Agriculture". *Monthly Review*.

[5] "Fidel's sustainable farmers". *The Economist*. 1999-04-24. Retrieved 17 September 2012.

[6] "The Castros, Cuba and America: On the road towards capitalism". *The Economist*. 2012-03-24. Retrieved 17 September 2012.

[7] Howard, April (2006). "How Green Is That Garden?". *E - The Environmental Magazine* (Earth Action Network, Inc.) **17**: 18–20. Retrieved 2010-05-18.

32.6   External links

- Urban Agriculture in Cuba (Photo Essay), Noah Friedman-Rudovsky, Oct 18 2012, NACLA.org

- "The Urban Agriculture of Havana," *Monthly Review*, 2009-Jan

- Case Study in Urban Agriculture: Organiponicos in Cienfuegos, Cuba

- Garden Activist: Cuba's Second Revolution

- The Growing Success of Organoponicos, Greenhouse Canada, by Gary Jones

- Changes on the Horizon for Cuba's Sustainable Agriculture

- Eat Local: Cuba's Urban Gardens Raise Food on Zero Emission

- Greg Morsbach Cuba's organic revolution BBC, June 27, 2001.

- Food Photography: Organic Agriculture in Cuba

- Bill McKibben The Cuba diet: What will you be eating when the revolution comes? *Harper's Magazine* April 1995.

- Esteban Israel In "eat local" movement, Cuba is years ahead Reuters, December 15, 2008.

- Andrew Buncombe The good life in Havana: Cuba's green revolution *The Independent* 8 August 2006

- Scott G. Chaplowe Havana's Popular Gardens: Sustainable Urban Agriculture, *WSAA Newsletter*, Fall 1996, Vol. 5, No. 22. Reprinted at cityfarmer.org

Chapter 33

Sebakh

Sebakh (less commonly transliterated as *sebbakh*) is an Aramaic word which translates to "dry land" in English. This term is used to described decomposed organic material that can be employed both as an agricultural fertilizer and as a fuel for fires.

33.1 Composition

Most sebakh consists of ancient, deteriorated mud brick. Mud brick was a primary building material in ancient Egypt. This material is composed of ancient mud mixed with the nitrous compost of the hay and stubble that the bricks were originally formulated with to give added strength before being baked in the sun.

33.2 Affecting archaeology

A common practice in Egypt, in the late nineteenth and early twentieth century, was for farmers to obtain government permits to remove this material from ancient mounds; such farmers were known as 'sebakhin'. Mounds indicating the location of ancient cities are also known as a *tell*, or *tel*.

An archaeological site could provide an excellent source of sebakh because decomposed organic debris creates a soil very rich in nitrogen. Nitrogen is an essential component in fertilizers used for plant crops.

Numerous potentially valuable archaeological finds were unfortunately destroyed by farmers in this way. However, sebakh digging also led to the discovery of archaeological finds that might otherwise have gone undetected.

33.3 Amarna

Sebakh is most commonly associated with the finding of the site of Amarna (Arabic: العمارنة al- ʿamārnä). In 1887, a local inhabitant who was delving into sebakh deposits accidentally discovered more than 300 cuneiform tablets that turned out to be Pharaonic records of correspondence. These tablet letters, known as the Amarna Letters, have provided much valuable historical and chronological data, as well as information bearing on Egyptian diplomatic relations with her neighbors at that time.

33.4 External links

- University of Southampton, 2002 - **Sebakh Excavations and the Written Material** (examples of sebakh diggings)

33.5   References

- Egyptology Online (sebakh used as fertilizer)

- Hierakonpolis Online (archaeological sebakh digging)

Chapter 34

Spent mushroom compost

Spent mushroom compost is the residual compost waste generated by the mushroom production industry. It is readily available (bagged, at nursery suppliers), and its formulation generally consists of a combination of wheat straw, dried blood, horse manure and ground chalk, composted together. It is an excellent source of humus, although much of its nitrogen content will have been used up by the composting and growing mushrooms. It remains, however, a good source of general nutrients (0.7% N, 0.3% P, 0.3% K plus a full range of trace elements), as well as a useful soil conditioner. However, due to its chalk content, it may be alkaline, and should not be used on acid-loving plants, nor should it be applied too frequently, as it will overly raise the soil's pH levels.*[1]

Mushroom compost may also contain pesticide residues, particularly organochlorides used against the fungus gnat. If the compost pile was stored outside, it may contain grubs or other insects attracted to decaying matter. Chemicals may also have been used to treat the straw, and also to sterilize the compost. Therefore, the organic gardener must be careful regarding the sourcing of mushroom compost; if in doubt, samples can be analyzed for contamination – in the UK, the Department for Environment, Food and Rural Affairs is able to advise regarding this issue.

Commercially available 'spent' mushroom compost is not always truly spent. It is sold by mushroom farms when it is no longer producing commercially viable yields of mushrooms. It can be used to grow further smaller crops of mushrooms before final use on the garden.

34.1 References

[1] Bradley, Steve (2004). *Vegetable Gardening: Growing and Harvesting Vegetables*. Murdoch Books. ISBN 1-74045-519-3.

34.2 External links

- How to grow mushrooms in spent mushroom compost before using in garden beds

Chapter 35

Stubble-mulching

Stubble-mulching refers to leaving the stubble (agriculture) or crop residue essentially in place on the land as a surface cover during a fallow period. Stubble-mulching can prevent erosion from wind or water and conserve soil moisture.

35.1   References

- Ⓩ This article incorporates public domain material from the Congressional Research Service document "Report for Congress: Agriculture: A Glossary of Terms, Programs, and Laws, 2005 Edition" by Jasper Womach.

Chapter 36

Used coffee grounds

Used coffee grounds in boxes.

Used coffee grounds are the waste product from brewing coffee. In the late 19th century, used coffee grounds were used to adulterate pure coffee.[*][1] In gardens, coffee grounds may be used for composting or as a mulch[*][2] as they are known to slowly release nitrogen into the soil. The coffee grounds are rich in potassium, magnesium and phosphorus. They are especially appreciated by worms and acid-loving plants such as blueberries.[*][3] Gardeners have reported the use of used coffee grounds as a slug and snail repellent,[*][2][*][4] but this has not yet been scientifically tested. Some commercial coffee shops run initiatives to prevent the grounds from going to waste, including Starbucks' "Grounds for your Garden" project,[*][5] and community sponsored initiatives exist, such as "Ground to Ground".[*][6]

Composting worms moving about in used coffee grounds.

Used coffee grounds have other homemade uses in wood staining, air fresheners, and body soap scrubs.[*][2] They may also be used industrially in biogas production or to treat wastewater.[*][7]

36.1   See also

- Ecological effects of coffee

36.2   Refences

[1] Pendergrast, Mark "*Uncommon grounds : the history of coffee and how it transformed our world*" 2010 Basic Books. ISBN 978-0-465-02404-9

[2] "Don't Throw Out Your Leftover Coffee Grounds!". Huffington Post. 4 August 2014. Retrieved 25 December 2014.

[3] Martin, Deborah L; Gershuny, Grace, eds. (1992). "Coffee wastes" . *The Rodale book of composting*. Emmaus, PA: Rodale Press. p. 86. ISBN 978-0-87857-991-4. Retrieved January 5, 2010.

[4] "NORTH COAST GARDENING: Winter vegetable growing" . Eureka Times-Standard. 24 December 2014. Retrieved 25 December 2014.

[5] "Grounds for Your Garden" . Starbucks.com. Retrieved October 26, 2011.

[6] "About Us | Coffee Grounds to Ground". Groundtoground.org. Retrieved October 26, 2011.

[7] Chalker-Scott, Ph.D, Linda (2009). "Coffee grounds—will they perk up plants?" (PDF). *Master Gardener*. Puyallup Research and Extension Center, Washington State University. Retrieved 25 December 2014.

Chapter 37

Uses of compost

Compost is a versatile product resulting from composting - the biodegradation of organic waste, industrially, commercially or domestically produced. Composting can be carried out at the household level, in garden composters or in composting toilets, or at municipal level at centralised composting plants. The method of producing the compost has an influence on its possible uses in terms of quantity and quality considerations.

The basic use of compost is conditioning and fertilizing soil by the addition of humus, nutrients and beneficial soil bacteria, with a wide range of specific applications.

37.1   Agriculture

On the open ground, for growing wheat, corn, soybeans, and similar crops, compost can be broadcast across the top of the soil using spreader trucks or spreaders pulled behind a tractor. It is expected that the spread layer is very thin (approximately 6 mm (0.25 in.)) and worked into the soil prior to planting. However, application rates of 25 mm (one in.) or more are not unusual when trying to rebuild poor soils or control erosion. Due to the extremely high cost of compost per unit of nutrients in the western world (such as USA) on-farm use is relatively rare since rates over 4 tons/acre can not be afforded. This is unfortunate and results from over-emphasis on "recycling organic matter" than on "sustainable nutrients". In other countries such as Germany, where compost distribution and spreading are partially subsidized in the original waste fees, compost is used more frequently on open ground, but only on the premise of nutrient "sustainability" *[1]

In plasticulture, strawberries, tomatoes, peppers, melons, and other fruits and vegetables are often grown under plastic to control temperature, retain moisture and control weeds. Compost may be banded (applied in strips along rows) and worked into the soil prior to bedding and planting, be applied at the same time the beds are constructed and plastic laid down, or used as a "top dressing".

Many crops are not seeded directly in the field but are started in seed trays in a greenhouse (see transplanting). When the seedlings reach a certain stage of growth, they are transplanted in the field. Compost can be used as an ingredient in the mix used to grow the seedlings, but is not normally used as the only planting substrate. The crop to be grown and the seeds' sensitivity to nutrients, salts, etc. dictates the ratio of the blend, and maturity is important to insure that oxygen deprivation will not occur or that no lingering phyto-toxins remain.*[2]

37.2   Horticulture

Compost is used in horticulture in a wide range of contexts. In raised bed gardening, compost can be mixed with sand, clay, aged sawdust, and other materials to create an enriched mix for landscape beds or raised-bed gardens. Compost should be no more than 30 percent of the total mix. Use a high quality mature compost to avoid nutrient and oxygen competition with plants.

In a container garden, as in bedding mixes, compost may be a beneficial ingredient in potting media, used up to 30 percent of the total mix, depending on salinity and maturity. It is considered a partial substitute for peat moss, but generally lacks the porosity and water-holding capacity of peat so must be used in limited percentages. The nutrient content of compost can also reduce the need for supplemental chemical fertilizers, although this has to be determined in each situation.

Excavated areas around the foundation of new buildings are backfilled when construction is complete, but these planting zones may contain rubble, residues of toxic chemicals, and other undesirable substances. Removing the backfill and replacing it with a soil/compost mix will improve soil structure and give foundation plantings a healthier start.

Two or more inches of compost can be used alone or in conjunction with conventional mulch products to keep root zones cool, conserve moisture, and act as a slow-release fertilizer, provided the product is course textured and mature. For a weed barrier, double or triple the depth of compost can be used, placed on top of a thick layer of newspapers, to replace geomembrane weed barriers. This is obviously only true if the compost is weed free; many are not.

For trees and shrubs, mixes of *well aged* compost with the native soils can be used as backfill. Immature composts may cause settling and young root disturbance due to oxygen deprivation. Seasonally, top dress with compost to the drip line and rake into the soil.

To establish new turf areas (lawns, recreation fields, golf courses), compost can be applied prior to seeding or sodding and work into the soil. Compost can seasonally be used to top dress and may also be raked into the soil. Some turf farms also use compost, growing grass in a couple of inches of the material to prevent topsoil loss.

37.3 Erosion control

Topsoil loss is a serious ecological issue. The use of compost to control sediment run-off and fight erosion is a relatively new technology, now being adopted by local authorities, developers, farmers, and other major disturbers of soil as another tool to reduce topsoil loss.

A layer of compost spread over a disturbed area of soil is called a compost blanket. With a high water-holding capacity, compost is not tilled into the soil but remains on the surface to temper the impact of rainfall. Even small amounts can help, but typical recommendations call for a 5 cm (2 in.) layer to insure adequate surface coverage. The blanket can also be directly planted into.

Compost berms and socks are used alone or in conjunction with compost blankets to mitigate the impact of high volume water discharges and flows. Compost berms are more aesthetically pleasing than silt fences and eliminate the need to remove the berm when the project is complete. Over time, a compost berm simply biodegrades and returns to the earth. As the name implies, a compost sock is a mesh tube stuffed with compost. Socks stand up better to heavy equipment, can be anchored in place, and are easily removed/reused. If a biodegradable fiber is used for the sock, it can also be left in place to biodegrade. This is rarely if ever practiced, however, since it defeats the idea of the sock.

37.4 Special uses

Additional special uses for compost include use as a planting media for constructed or artificial wetlands, as a cap for a landfill cell when it is closed to encourage vegetation and reduce erosion, and as erosion control along streambanks to restore functionality and beauty to riparian zones while possibly mitigating future damage.

37.5 Regulation and voluntary standards

EPA Class A and B guidelines in the U.S.A.[3] were developed solely to manage the processing and beneficial reuse of sludge, also now called biosolids, following the US EPA ban of ocean dumping. About 26 American states now require composts to be processed according to these federal protocols for pathogen and vector control, even though the application to non-sludge materials has not been scientifically tested. An example is that green waste composts are used at much higher rates than sludge composts were ever anticipated to be applied at.*[4] U.K guidelines also exist regarding compost quality,*[5] as well as Canadian,*[6] Australian,*[7] and the various European states.*[8]

In the USA some compost manufacturers participate in a testing program offered by a private lobbying organization called the U.S. Composting Council. The USCC was originally established in 1991 by Procter & Gamble to promote composting of disposable diapers, following state mandates to ban diapers in landfills, which caused a national uproar. Ultimately the idea of composting diapers was abandoned, partly since it was not proven scientifically to be possible, and mostly because the concept was a marketing stunt in the first place. After this, composting emphasis shifted back to recycling organic wastes previously destined for landfills. There are no bonafide quality standards in America, but the USCC sells a seal called "Seal of Testing Assurance" *[9] (also called "STA"). For a considerable fee, the applicant may display the USCC logo on products, agreeing to volunteer to customers a current laboratory analysis that includes parameters such as nutrients, respiration rate, salt content, pH, and limited other indicators.*[10] However, the STA program is not ISO approved, and is a financially beneficial activity for the private USCC, an organization that does disclose its books (in 2009 USCC earned $65,000 from STA fees). Some argue that the existence of STA means EPA or USDA do not have to regulate composts.

37.6   References

[1] http://www.landwirtschaft-mlr.baden-wuerttemberg.de/servlet/PB/show/1118971/Landinfo_Nachhaltige%20Kompostanwendung%20in%20der%20Landwirtschaft-%20Ergebnisse%20eines%20mehrj%E4hrigen%20DBU-Projektes%20aus%20Baden-W%FCrttemberg.pdf

[2] Phytotoxicity and maturation

[3] EPA Class A standards

[4] EPA regulations for compost use

[5] British Standards Institute Specifications

[6] Consensus Canadian national standards

[7] Australian quality standards

[8] EU member compost standards

[9]

[10] US Composting Council testing parameters

37.7   External links

- Compost Tea

- Discussion of world-wide compost standards

- Provincial composting regulation in Canada

- Seal of Testing Assurance

- Article on compost use for sediment and erosion control from the Carolinas Composting Council.

- Article on compost use for sediment and erosion control from the University of Georgia Cooperative Extension Service.

- EPA paper on anaerobic compost-constructed wetlands

Chapter 38

Vermicompost

Vermicompost is the product or process of composting using various worms, usually red wigglers, white worms, and other earthworms to create a heterogeneous mixture of decomposing vegetable or food waste, bedding materials, and vermicast. Vermicast, also called worm castings, worm humus or worm manure, is the end-product of the breakdown of organic matter by an earthworm.[1] These castings have been shown to contain reduced levels of contaminants and a higher saturation of nutrients than do organic materials before vermicomposting.[2]

Containing water-soluble nutrients, vermicompost is an excellent, nutrient-rich organic fertilizer and soil conditioner.[3] This process of producing vermicompost is called *vermicomposting*.

38.1 Suitable species

One of the earthworm species most often used for composting is the Red Wiggler (*Eisenia fetida* or *Eisenia andrei*); *Lumbricus rubellus* (a.k.a. red earthworm or dilong (China)) is another breed of worm that can be used, but it does not adapt as well to the shallow compost bin as does *Eisenia fetida*. European nightcrawlers (*Eisenia hortensis*) may also be used. Users refer to European nightcrawlers by a variety of other names, including dendrobaenas, dendras, and Belgian nightcrawlers. African Nightcrawlers (*Eudrilus eugeniae*) are another set of popular composters. *Lumbricus terrestris* (a.k.a. Canadian nightcrawlers (US) or common earthworm (UK)) are not recommended, as they burrow deeper than most compost bins can accommodate.[4]

Blueworms (*Perionyx excavatus*) may be used in the tropics.[5]

These species commonly are found in organic-rich soils throughout Europe and North America and live in rotting vegetation, compost, and manure piles. They may be an invasive species in some areas.[1][6] As they are shallow-dwelling and feed on decomposing plant matter in the soil, they adapt easily to living on food or plant waste in the confines of a worm bin.

Composting worms are available to order online, from nursery mail-order suppliers or angling shops where they are sold as bait. They can also be collected from compost and manure piles. These species are not the same worms that are found in ordinary soil or on pavement when the soil is flooded by water.

38.2 Large scale

Large-scale vermicomposting is practiced in Canada, Italy, Japan, Malaysia, the Philippines, and the United States.[7][8] The vermicompost may be used for farming, landscaping, to create compost tea, or for sale. Some of these operations produce worms for bait and/or home vermicomposting.

There are two main methods of large-scale vermiculture. Some systems use a windrow, which consists of bedding materials for the earthworms to live in and acts as a large bin; organic material is added to it. Although the windrow has no

Rotary screen harvested vermicompost, composed of worm castings

physical barriers to prevent worms from escaping, in theory they should not due to an abundance of organic matter for them to feed on. Often windrows are used on a concrete surface to prevent predators from gaining access to the worm population.

The second type of large-scale vermicomposting system is the raised bed or flow-through system. Here the worms are fed an inch of "worm chow" across the top of the bed, and an inch of castings are harvested from below by pulling a

Direction of wave motion in wormbed

Feed is always added to this side of the bed. After mulching and continuous feeding, this side of the bed becomes larger and pushes across.

The backside of the bed, rich with castings, can be removed for sale. This side is prone to weeds and aggressive grasses.

Movement of castings through a worm bed.

breaker bar across the large mesh screen which forms the base of the bed.

Because red worms are surface dwellers constantly moving towards the new food source, the flow-through system eliminates the need to separate worms from the castings before packaging. Flow-through systems are well suited to indoor facilities, making them the preferred choice for operations in colder climates.

38.3   Small scale

For vermicomposting at home, a large variety of bins are commercially available, or a variety of adapted containers may be used. They may be made of old plastic containers, wood, Styrofoam, or metal containers. The design of a small bin usually depends on where an individual wishes to store the bin and how they wish to feed the worms.

Some materials are less desirable than others in worm bin construction. Metal containers often conduct heat too readily, are prone to rusting, and may release heavy metals into the vermicompost. Some cedars, Yellow cedar, and Redwood contain resinous oils that may harm worms,*[9] although Western Red Cedar has excellent longevity in composting conditions. Hemlock is another inexpensive and fairly rot-resistant wood species that may be used to build worm bins.*[10]

Bins need holes or mesh for aeration. Some people add a spout or holes in the bottom for excess liquid to drain into a tray for collection. Worm compost bins made from plastic are ideal, but require more drainage than wooden ones because they are non-absorbent. However, wooden bins will eventually decay and need to be replaced.

Small-scale vermicomposting is well-suited to turn kitchen waste into high-quality soil amendments, where space is limited. Worms can decompose organic matter without the additional human physical effort (turning the bin) that bin composting requires.

Composting worms which are detritivorous (eaters of trash), such as the red wiggler *Eisenia fetidae*, are epigeic (surface dwellers) together with symbiotic associated microbes are the ideal vectors for decomposing food waste. Common earthworms such as *Lumbricus terrestris* are anecic(deep burrowing) species and hence unsuitable for use in a closed system.*[11] Other soil species that contribute include insects, other worms and molds.*[12]

38.4   Climate and temperature

The most common worms used in composting systems, redworms (*Eisenia foetida, Eisenia andrei,* and *Lumbricus rubellus*) feed most rapidly at temperatures of 15–25 °C (59-77 °F). They can survive at 10 °C (50 °F). Temperatures above 30 °C (86 °F) may harm them.*[13] This temperature range means that indoor vermicomposting with redworms is possible in

Demonstration home scale worm bin at a community garden site - painted plywood

all but tropical climates.[*][14] (Other worms like Perionyx excavatus are suitable for warmer climates.[*][15]) If a worm bin is kept outside, it should be placed in a sheltered position away from direct sunlight and insulated against frost in winter.

It is necessary to monitor the temperatures of large-scale bin systems (which can have high heat-retentive properties), as the feedstocks used can compost, heating up the worm bins as they decay and killing the worms.

38.5   Feedstock

There are few food wastes that vermicomposting cannot compost, although meat waste and dairy products are likely to putrefy, and in outdoor bins can attract vermin. Green waste should be added in moderation to avoid heating the bin.

38.5.1   Small-scale or home systems

Such systems usually use kitchen and garden waste, using "earthworms and other microorganisms to digest organic wastes, such as kitchen scraps".[*][16] This includes:

- Fruits and vegetables (not including citrus, other "high acid" foods, onions and garlic)

- Vegetable and fruit peels and ends

- Coffee grounds and filters

- Loose tea and whole bags (even those with high tannin levels)

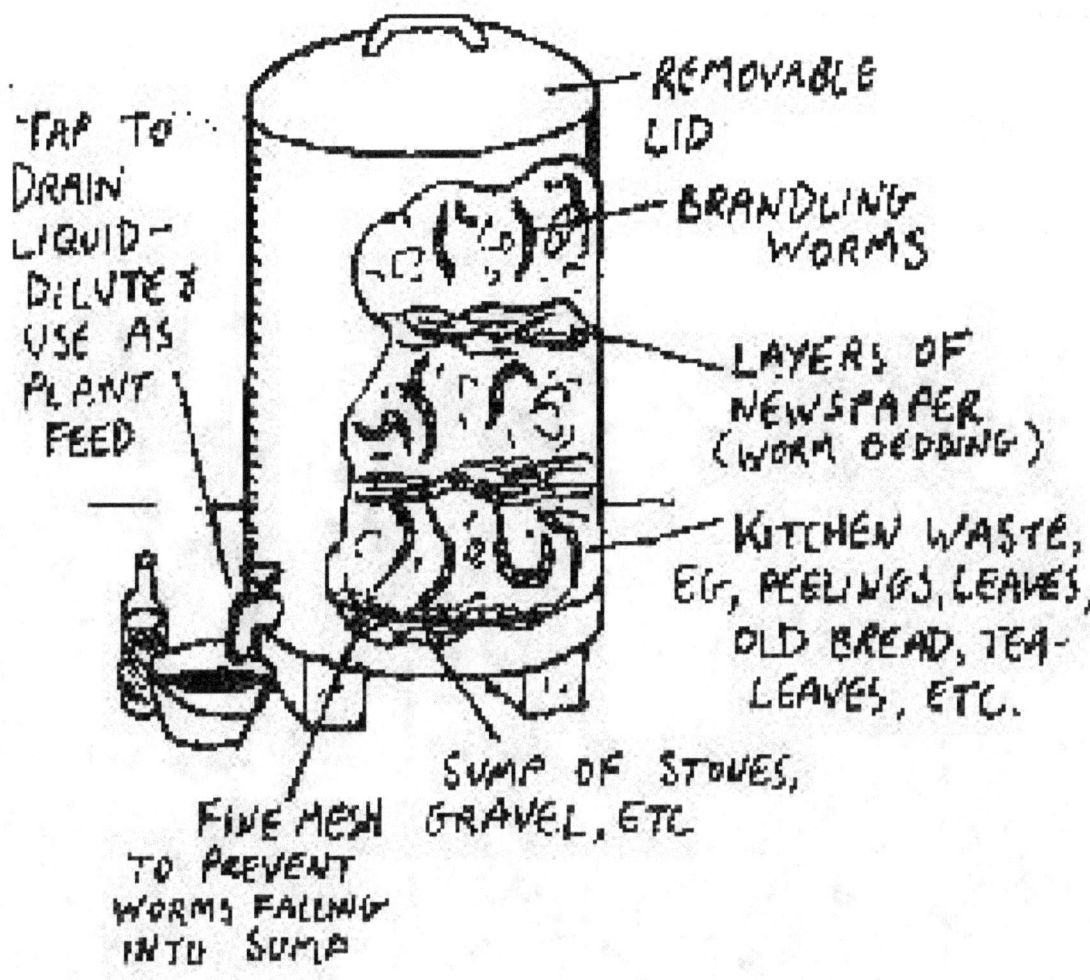

Diagram of a household-scale worm composting bin

- Grains such as bread, cracker and cereal (including moldy and stale, but not including anything oily, such as pizza crusts)

- Ground up eggshells (well rinsed)

- Leaves and plant clippings (not sprayed with pesticides[17] and no evergreen)

38.5.2   Large-scale or commercial

Such vermicomposting systems need reliable sources of large quantities of food. Systems presently operating[18] use:

- Dairy cow or pig manure

- Sewage sludge.[19][20]

- Brewery waste

- Cotton mill waste

- Agricultural waste

- Food processing and grocery waste

- Cafeteria waste

- Grass clippings and wood chips

38.6   Harvesting

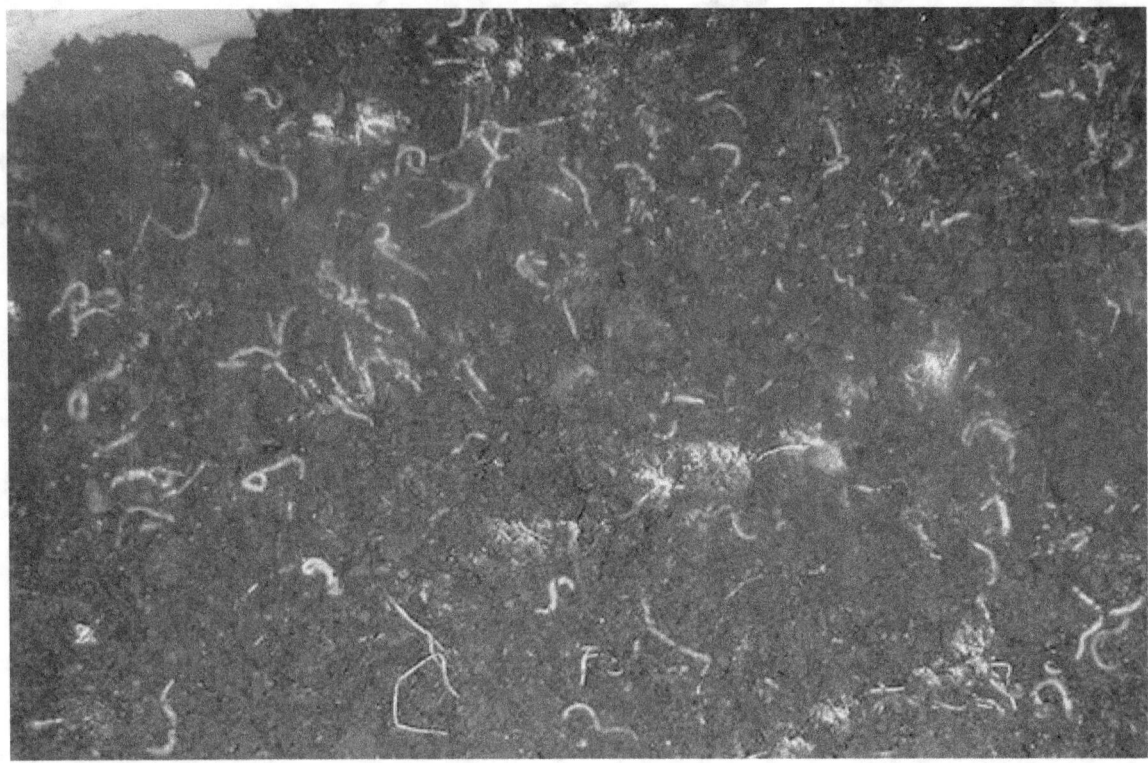

Worms in a bin being harvested

Vermicompost is ready for harvest when it contains few-to-no scraps of uneaten food or bedding . There are several methods of harvesting from small-scale systems: "dump and hand sort" , "let the worms do the sorting" , "alternate containers" and "divide and dump." [21] These differ on the amount of time and labor involved and whether the vermicomposter wants to save as many worms as possible from being trapped in the harvested compost.

While harvesting, it's also a good idea to try to pick out as many eggs/cocoons as possible and return them to the bin. Eggs are small, lemon-shaped yellowish objects that can usually be seen pretty easily with the naked eye and picked out. [22]

38.7   Properties

Vermicompost has been shown to be richer in many nutrients than compost produced by other composting methods. [23] It has also outperformed a commercial plant medium with nutrients added, but levels of magnesium required adjustment, as did pH. [24]

However, in one study it has been found that homemade backyard vermicompost was lower in microbial biomass, soil microbial activity, and yield of a species of ryegrass[25]than municipal compost,[25]

It is rich in microbial life which converts nutrients already present in the soil into plant-available forms.

Unlike other compost, worm castings also contain worm mucus which helps prevent nutrients from washing away with the first watering and holds moisture better than plain soil.*[26]

38.8 Benefits

Soil

- Improves soil aeration

- Enriches soil with micro-organisms (adding enzymes such as phosphatase and cellulase)

- Microbial activity in worm castings is 10 to 20 times higher than in the soil and organic matter that the worm ingests *[27]

- Attracts deep-burrowing earthworms already present in the soil

- Improves water holding capacity*[28]

Plant growth

- Enhances germination, plant growth, and crop yield

- Improves root growth and structure

- Enriches soil with micro-organisms (adding plant hormones such as auxins and gibberellic acid)

Economic

- Biowastes conversion reduces waste flow to landfills

- Elimination of biowastes from the waste stream reduces contamination of other recyclables collected in a single bin (a common problem in communities practicing single-stream recycling*[29])

- Creates low-skill jobs at local level

- Low capital investment and relatively simple technologies make vermicomposting practical for less-developed agricultural regions

Environmental

- Helps to close the "metabolic gap" through recycling waste on-site

- Large systems often use temperature control and mechanized harvesting, however other equipment is relatively simple and does not wear out quickly

- Production reduces greenhouse gas emissions such as methane and nitric oxide (produced in landfills or incinerators when not composted or through methane harvest)*[30]

Mid-scale worm bin (1 m X 2.5 m up to 1 m deep), freshly refilled with bedding

38.9   As fertilizer

Vermicompost can be mixed directly into the soil, or steeped in water and made into a worm tea by mixing some vermicompost in water, bubbling in oxygen with a small air pump, and steeping for a number of hours or days.

The microbial activity of the compost is greater if it is aerated during this period. The resulting liquid is used as a fertilizer or sprayed on the plants.

The dark brown waste liquid, or leachate, that drains into the bottom of some vermicomposting systems as water-rich foods break down, is best applied back to the bin when added moisture is needed due to the possibility of phytotoxin content and organic acids that may be toxic to plants.*[9]

The pH, nutrient, and microbial content of these fertilizers varies upon the inputs fed to worms. Pulverized limestone, or calcium carbonate can be added to the system to raise the pH.

38.10   Troubleshooting

38.10.1   Smells

When closed, a well-maintained bin is odorless; when opened, it should have little smell if any smell is present, it is earthy.*[31] Worms require gaseous oxygen.*[32] Oxygen can be provided by airholes in the bin, occasional stirring of bin contents, and removal of some bin contents if they become too deep or too wet. If decomposition becomes anaerobic from excess wet feedstock added to the bin, or the layers of food waste have become too deep, the bin will begin to smell of ammonia.

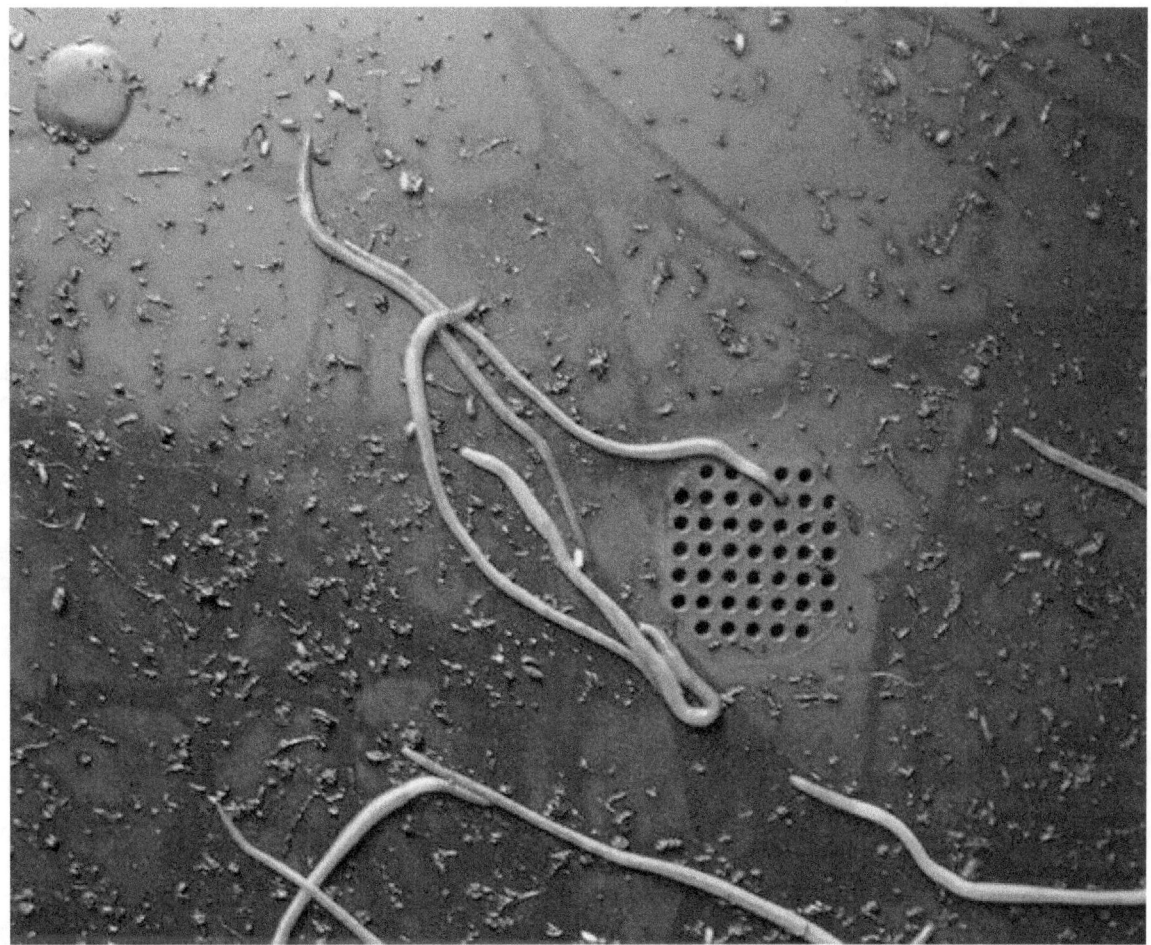

Worms and fruit fly pupas under the lid of a home worm bin.

38.10.2 Moisture

If decomposition has become anaerobic, to restore healthy conditions and prevent the worms from dying, the smelly, excess waste water must be removed and the bin returned to a normal moisture level. To do this, first reduce addition of food scraps with a high moisture content and second, add fresh, dry bedding such as shredded newspaper to your bin, mixing it in well.

38.10.3 Pest species

Pests such as rodents and flies are attracted by certain materials and odors, usually from large amounts of kitchen waste, particularly meat. Eliminating the use of meat or dairy product in a worm bin decreases the possibility of pests.*[33]

In warm weather, fruit and vinegar flies breed in the bins if fruit and vegetable waste is not thoroughly covered with bedding. This problem can be avoided by thoroughly covering the waste by at least 2 inches (5.1 cm) of bedding. Maintaining the correct pH (close to neutral) and water content of the bin (just enough water where squeezed bedding drips a couple of drops) can help avoid these pests as well.

38.10.4   Worms escaping

Worms generally stay in the bin, but may try to leave the bin when first introduced, or often after a rainstorm when outside humidity is high.*[34] Maintaining adequate conditions in the worm bin and putting a light over the bin when first introducing worms should eliminate this problem.*[35]

38.10.5   Nutrient levels

Commercial vermicomposters test, and may amend their products to produce consistent quality and results. Because the small-scale and home systems use a varied mix of feedstocks, the nitrogen, potassium and phosphorus content of the resulting vermicompost will also be inconsistent. NPK testing may be helpful before the vermicompost or tea is applied to the garden.

In order to avoid over-fertilization issues, such as nitrogen burn, vermicompost can be diluted as a tea 50:50 with water, or as a solid can be mixed in 50:50 with potting soil.*[36]

The mucus produced creates a natural time-release fertilizer which cannot burn plants.*[37]

38.11   See also

- Composting

- Decompiculture

- Fertilizer

- Home composting

- Vermiponics, use of wormbin leachate in hydroponics

- Waste management

38.12   Notes

[1]   "Paper on Invasive European Worms" . Retrieved 2009-02-22.

[2]   Ndegwa, P.M.; Thompson, S.A.; Das, K.C. (1998).   "Effects of stocking density and feeding rate on vermicomposting of biosolids" (PDF). *Bioresource Technology* **71**: 5–12. doi:10.1016/S0960-8524(99)00055-3.

[3]   Coyne, Kelly and Erik Knutzen. *The Urban Homestead: Your Guide to Self-Sufficient Living in the Heart of the City.* Port Townsend: Process Self Reliance Series, 2008.

[4]   "Composting with earthworms" . Herron Farms Dawsonville Ga. Retrieved March 26, 2013.

[5]   "Composting Worms for Hawaii" (PDF). Retrieved 2009-02-22.

[6]   "Great Lakes Worm Watch" . Retrieved 2009-02-22.

[7]   "Vermicomposting: A Better Option for Organic Solid Waste Management" (PDF). Retrieved 2009-02-21.

[8]   "Compost Tea" . Retrieved 2009-02-22.

[9]   "Raising Earthworms Successfully" (PDF). Retrieved 2009-03-04.

[10]   Archived July 24, 2010 at the Wayback Machine

[11]   "The Worm Dictionary and Vermiculture Reference Center" . Working Worms. Retrieved 3 October 2012.

[12] Trautmann, Nancy. "Invertebrates of the Compost Pile". Cornell Center for the Environment. Retrieved 2012-10-03.

[13] Appelhof, p. 3

[14] "Map of vermicomposters". Vermicomposters.com. Retrieved 2012-10-03.

[15] Appelhof, p. 41

[16] Selden, Piper; DuPonte, Michael; Sipes, Brent; Dinges, Kelly (August 2005). "Small-Scale Vermicomposting" (PDF). *Home Garden* (University of Hawai'i) **45**. Retrieved 2012-10-03.

[17] Reinecke, SA; Reinecke, AJ (February 2007). "The impact of organophosphate pesticides in orchards on earthworms in the Western Cape, South Africa." (PDF). *Ecotoxicology and Environmental Safety* **66** (2): 244–51. doi:10.1016/j.ecoenv.2005.10.006. PMID 16318873.

[18] Latest Developments In Mid-To-Large-Scale Vermicomposting Archived February 10, 2015 at the Wayback Machine

[19] Archived October 3, 2009 at the Wayback Machine

[20] Lotzof, M. "Very Large Scale Vermiculture in Sludge Stabilisation". Vermitech Pty Limited. Retrieved 2012-10-03.

[21] Appelhof, pp. 79-86

[22]

[23] Dickerson, George W. (June 2001). "Vermicomposting: Guide H-164" (PDF). New Mexico State University. Retrieved 2012-10-03.

[24] Sherman, Rhonda. "Earthworm Castings as Plant Growth Media". Department of Biological and Agricultural Engineering at NCSU. Retrieved 2012-10-03.

[25] Lazcano, Cristina; Gómez-Brandón, María; Domínguez, Jorge (July 2008). "Comparison of the effectiveness of composting and vermicomposting for the biological stabilization of cattle manure" (PDF). *Chemosphere* **72** (7): 1013–1019. doi:10.1016/j.chemosphere.2008.04.016.

[26] Nancarrow, Loren; Taylor, Janet Hogan (1998). *The Worm Book: The Complete Guide to Gardening and Composting with Worms* Ten Speed Press. p. 4. ISBN 978-0-89815-994-3.

[27] Logsdon, Gene (October 1994). "Worldwide progress in vermicomposting". *BioCycle* **35** (10): 63.

[28] Appelhof, p. 111

[29] See Wikipedia article on single-stream recycling.

[30] "Waste Management to tap landfill methane". MSNBC. June 27, 2007. Retrieved 2012-10-03.

[31] Appelhof, p. 113

[32] Appelhof, p. 92

[33] "Manual of On-Farm Vermicomposting and Vermiculture" (PDF). p. 8. Retrieved 2009-12-10.

[34] Compost Worm Escape

[35]

[36] Grant, Tim; Littlejohn, Gail (2004). *Teaching Green, The Middle Years.* Gabriola Island, B.C.: New Society Publishers. p. 121. ISBN 978-0-86571-501-1.

[37] "Compost or Worm Castings?". VermiDirt. Retrieved 2012-10-03.

38.13 References

- Appelhof, Mary (2007). *Worms Eat My Garbage* (2nd ed.). Kalamazoo, Mich.: Flowerfield Enterprises. ISBN 978-0-9778045-1-1.

38.14   External links

- Learning materials related to Vermicompost at Wikiversity

Chapter 39

Windrow composting

Windrow turner used on maturing piles at a biosolids composting facility in Canada.

In agriculture, **windrow composting** is the production of compost by piling organic matter or biodegradable waste, such as animal manure and crop residues, in long rows (*windrows*). This method is suited to producing large volumes of compost. These rows are generally turned to improve porosity and oxygen content, mix in or remove moisture, and redistribute cooler and hotter portions of the pile. Windrow composting is a commonly used farm scale composting method. Composting process control parameters include the initial ratios of carbon and nitrogen rich materials, the amount of bulking agent added to assure air porosity, the pile size, moisture content, and turning frequency.

The temperature of the windrows must be measured and logged constantly to determine the optimum time to turn them for quicker compost production.

Maturing windrows at an in-vessel composting facility.

39.1   Compost windrow turners

Compost windrow turners were developed to produce compost on a large scale by Fletcher Sims Jr. of Canyon, Texas . They are traditionally a large machine that straddles a windrow of 4 feet (1.25 meters) or more high, by as much as 12 feet (3.5 meters) across. Although smaller machines exist for small windrows, most operations use large machines for volume production. Turners drive through the windrow at a slow rate of forward movement. They have a steel drum with paddles that are rapidly turning. As the turner moves through the windrow, fresh air (oxygen) is injected into the compost by the drum/paddle assembly, and waste gases produced by bacterial decomposition are vented. The oxygen feeds the aerobic bacteria and thus speeds the composting process.

Utilization

To properly use a compost windrow turner, it is ideal to compost on a hard surfaced pad. Heavy-duty compost windrow turners allow the user to obtain optimum results with the aerobic hot composting process. By using four wheel drive or tracks the windrow turner is capable of turning compost in windrows located in remote locations. With a self-trailering option this allows the compost windrow turner to convert itself into a trailer to be pulled by a semi-truck tractor. These two options combined allow the compost windrow turner to be easily hauled anywhere and to work compost windrows in muddy and wet locations.

39.2   Specific applications

Molasses-based distilleries all over the world generate large amount of effluent termed as spent wash or vinasse. For each liter of alcohol produced, around 8 liters of effluent is generated. This effluent has COD of 1,50,000 PPM and BOD of 60,000 PPM and even more. This effluent needs to be treated and the only effective method for conclusive disposal is by composting.

Sugar factories generate pressmud / cachaza during the process and the same has about 30% fibers as carbon and has large amounts of water. This pressmud is dumped on prepared land in the form of 100 meters long windrows of 3 meters x

1.5 meters and spent wash is sprayed on the windrow while the windrow is being turned. These machines help consume spent wash of about 2.5 times of the volume of the pressmud, which means that a 100 meters of windrow accommodates about 166 MT of pressmud and uses about 415 m^3 of Spent wash in 50 days.

Microbial Culture(organic solution) TRIO COM-CULT is used about 1 kg per MT of pressmud for fast de-composing of the spent wash. Hundreds of thousands of square meters of spent wash is being composted all over the world in countries like India, Colombia, Brazil, Thailand, Indonesia, South Africa etc.

The compost yard has to be prepared in such a way that the land is impervious and does not allow the liquid effluent to pass down into the earth. The compost thus generated is of excellent quality and is rich in nutrients.

39.3   See also

- Aerated static pile composting

- In-vessel composting

39.4   References

- Coufal, Dr, Craig (2008). "In-House Windrow Composting Q and A". Production Management Featured Articles Web site. Retrieved August 27, 2009.

39.5   External links

- Windrow Dynamics

- Glossary Composting

- Windrow Composting - Grass Straw (retrieved March 17, 2009)

39.6   Text and image sources, contributors, and licenses

39.6.1   Text

- **Bioeffector** *Source:* https://en.wikipedia.org/wiki/Bioeffector?oldid=628298945 *Contributors:* DragonflySixtyseven, Wavelength, Raupp, WOSlinker, Ironholds, FrescoBot and EmausBot

- **Compost** *Source:* https://en.wikipedia.org/wiki/Compost?oldid=682692287 *Contributors:* Eloquence, Bryan Derksen, -- April, Andre Engels, Jpsturm, Josh Grosse, Anthere, Zoe, Heron, Quercusrobur, Edward, Pit~enwiki, Menchi, Qaz, Paul A, Egil, DavidWBrooks, Stan Shebs, Ronz, Kokamomi, 5ko, Julesd, Glenn, Llull, º¡º, Raven in Orbit, Jengod, Dcoetzee, Ike9898, Jose Ramos, Samsara, Geraki, Mignon~enwiki, Pollinator, UninvitedCompany, Robbot, Moncrief, Goethean, Thunderbolt16, KellyCoinGuy, Antonin~enwiki, Tsavage, Pengo, Alan Liefting, MPF, Elf, Timpo, Bradeos Graphon, Guanaco, Solipsist, Bobblewik, LennartBolks~enwiki, JRR Trollkien, Chowbok, Beland, Madmagic, OverlordQ, Zantolak, Jeremykemp, Ukexpat, Sonett72, Stevenmattern, Discospinster, Vsmith, LindsayH, ESkog, Pedant, Tompw, Mister-Sheik, *drew, Kwamikagami, EurekaLott, CDN99, Bobo192, Spalding, Fir0002, Duk, Vortexrealm, Cohesion, Giraffedata, Kjkolb, Jeodesic, MPerel, Batneil, Frank101, Alansohn, Anthony Appleyard, Arthena, Paleorthid, Keflavich, Rohirok, Bart133, Velella, Mikeo, Kusma, Gene Nygaard, Dismas, Brookie, Abanima, Velho, Lincspoacher, Doctor Boogaloo, Uncle G, WadeSimMiser, Jwanders, Bennetto, Mandarax, Lego872, Kissekatt, Rjwilmsi, Valentinejoesmith, Harro5, HappyCamper, FlaBot, Eubot, AdnanSa, Latka, Margosbot~enwiki, Angstrom, EronMain, Benanhalt, Ahpook, YurikBot, Wavelength, RobotE, Cheesewire, Sceptre, Kollision, Carllindstrom, Splash, SpuriousQ, Stephenb, GeeJo, Leighblackall, NawlinWiki, Grandad, Nirvana2013, R'son-W, Brandon, Diotti, My Cat inn, Semperf, Natkeeran, DeadEyeArrow, Supspirit, CLW, Wknight94, Closedmouth, Xaxafrad, ArielGold, Whouk, Mejor Los Indios, ChemGardener, Erik Sandberg, SmackBot, Evilsofcompost, Rex the first, Chairman S., WookieInHeat, Edgar181, Cazort, Jonobennett, Gilliam, Donama, Ghosts&empties, Mycota, Andy M. Wang, Chris the speller, Bluebot, Persian Poet Gal, Salvor, Deli nk, ArcaneMachine, Brimba, Rrburke, Nahum Reduta, Nakon, 4hodmt, Drphilharmonic, DMacks, Torst, Byelf2007, Rory096, AThing, Srikeit, Euchiasmus, Gobonobo, Tim Q. Wells, Mr. Lefty, IronGargoyle, FrostyBytes, Werdan7, Metao, Peter Horn, ASVP, KJS77, Hu12, Informedbanker, BranStark, Natronomonas, Iridescent, Russcohn, Tony Fox, Dp462090, Courcelles, Ziusudra, Poolkris, JForget, Sakowski, Makeemlighter, DeLarge, Harej bot, LLucas, DoranM, Funnyfarmofdoom, Fnlayson, Peripitus, Yan24, Lugnuts, Hibou8, Tawkerbot4, Teratornis, Ameliorate!, Gimmetrow, Satori Son, Epbr123, Nahaboy, Marek69, Zé da Silva, Grayshi, Orfen, Dajagr, Derzsi Elekes Andor, Panphage, MadeHere, Jj137, TimVickers, Paul Christensen, Smartse, Gundam07th, Danger, Kent Witham, ScottM84, Kariteh, .alyn.post., JAnDbot, Deflective, Husond, Od1n, DuncanHill, Hut 8.5, TheEditrix2, Magioladitis, Bongwarrior, VoABot II, Yyyikes, CTF83!, Sustainableyes, MartinBot, Arjun01, CommonsDelinker, Artaxiad, J.delanoy, Pharaoh of the Wizards, Rlsheehan, Bogey97, Eliz81, Naniwako, Vanished user g454XxNpUVWvxzlr, DadaNeem, Ontarioboy, Brendan19, FuegoFish, Fiona C Mackenzie, Sjforman, DASonnenfeld, Idioma-bot, Signalhead, Sporti, Jeff G., W. B. Wilson, Barneca, JBazuzi, Philip Trueman, TXiKiBoT, AgamemnonZ, Joe2832, ElinorD, Aymatth2, Corvus cornix, LeaveSleaves, BotKung, Anarchangel, Scabether, Temporaluser, Gnowk, Red58bill, MrChupon, ZBrannigan, Demize, Sauronjim, L32007, SieBot, YonaBot, Yintan, Tttools123, Keilana, Jojalozzo, Robotchampion, OsamaBinLogin, Macnabr, Wmpearl, Antonio Lopez, Emo joe, Bingbing154, Rdoiron1, Hobartimus, Nancy, Dillard421, MitchCallison, Gardenmandy, Macdaddyfolife, Nn123645, Denisarona, ClueBot, GorillaWarfare, Dobermanji, The Thing That Should Not Be, Wysprgr2005, LisaSmegal, Connect 321, Uncle Milty, CounterVandalismBot, Auntof6, Tadhussey, Jcjenkins, OrBot, Excirial, Mindcry, Grey Matter, Fradol, Micha, NJGW, Schwellungswasser, MasterOfHisOwnDomain, Methanus, Ortos12, XLinkBot, Gnowor, Lumenos, Avoided, Gazimoff, Eroeben, Dkp205w, Apmorrison1993, Jojhutton, Zellfaze, MrOllie, CarsracBot, Bassbonerocks, FCSundae, Arben1951, Splodgeness, Cammos, Loser4132, 102orion, Tide rolls, Totorotroll, Jarble, Martin Hanson, Legobot, Yobot, עידן ד,2D, Fraggle81, Fmrauch, THEN WHO WAS PHONE?, Bmwjackson, Dmarquard, BruceMcAdam, AnomieBOT, DemocraticLuntz, 1exec1, Leangreenhome, Jim1138, Golb12, Jan Complace, Goodtimber, Materialscientist, Efefher, Puddypie007, Neurolysis, H9e3k80, Acuares, Apothecia, Wperdue, Jacquibeeadams, Twirligig, Adus123, Memepope313, RibotBOT, Mattg82, Rickproser, Wpnoone, Shadowjams, Noclock, FrescoBot, Lothar von Richthofen, Recognizance, Sickymicky, Winterst, Pinethicket, I dream of horses, Rule 56, Tyman510, Jschnur, Jorden12, FoxBot, Sintau.tayua, Siltloam, Lotje, Wien12, Japsmp, MrX, AbeColey, DARTH SIDIOUS 2, RjwilmsiBot, Tomlauter, DASHBot, John of Reading, Look2See1, Dewritech, RenamedUser01302013, Tommy2010, Winner 42, Mmeijeri, K6ka, Hollylogan, Azecha, Bamyers99, Dr Black Knife, Benjaminoakes, Jscoop5, FloGreen, Mlemacio, Inka 888, BioPupil, Hyronimus299, Agungsuko, VictorianMutant, CharlieEchoTango, Allisonlhjack, Anita5192, Petrb, Will Beback Auto, ClueBot NG, Peter James, Silly popper dude, Gilderien, Stephen.upton, Chester Markel, Mukitil, Tideflat, Cchanak, Nikkijean, Widr, Antiqueight, Avidgardener711, Ckgurney, Helpful Pixie Bot, Micklan, Gob Lofa, Gauravjuvekar, Bofum, BG19bot, Xamnidar, LioRelations, TopDog5450, Northamerica1000, Ealison, Fahdaftab, Badon, Erickthecompostguy, Corteli9, Volcomkewl, Snow Blizzard, BokashiGirl, ChrisBalch, Rudork, Cclehnen, Whitehousee, Palmfarm, EuroCarGT, Whynot777, Yobwej666, OscarK878, Vamfun, Teighbouy, Lugia2453, Frosty, Graphium, Jochen Burghardt, TheRealWallyrus, Corn cheese, Stevethor, TheBigSnax, I am One of Many, Howicus, Blogmaster121, Federales, Ugog Nizdast, Manul, FinallyAUsernameICanUse, Wafflesandcake, Actinomyces, Lagoset, Monkbot, Wolf.KT, Madhattertea, Justin Schaeffer, Madhav Jolly, Thaliaelias, GLG GLG, EvMsmile, KyleCMSmith, Capt.lingard, JMWt, KasparBot, Rm1911, Martaford, Degglenause and Anonymous: 570

- **Hügelkultur** *Source:* https://en.wikipedia.org/wiki/H%C3%BCgelkultur?oldid=680084542 *Contributors:* Jengod, Jjron, Jaerik, Cyfal, Yobot, Lothar von Richthofen, MrX, MusikAnimal, Lagoset, Ephemeralcas, EvMsmile and Anonymous: 2

- **List of composting systems** *Source:* https://en.wikipedia.org/wiki/List_of_composting_systems?oldid=646563429 *Contributors:* Vortexrealm, Vegaswikian, Amakuha, ChemGardener, Juandev, Vrac, Red58bill, Doctorfluffy, Axiomatica, Μάριος Ζηντίλης, Northamerica1000, EvMsmile and Anonymous: 4

- **Aerated static pile composting** *Source:* https://en.wikipedia.org/wiki/Aerated_static_pile_composting?oldid=660676288 *Contributors:* Skysmith, Alan Liefting, Vortexrealm, RJFJR, Rjwilmsi, Wavelength, Jlittlet, Alynna Kasmira, JohJak2, BirgitteSB, SmackBot, Miljoshi, Smsaladi, LLucas, Alphachimpbot, Walter Hartmann, Red58bill, Dthomsen8, Yobot, Duboshi, Northamerica1000 and Anonymous: 3

- **An Agricultural Testament** *Source:* https://en.wikipedia.org/wiki/An_Agricultural_Testament?oldid=628359164 *Contributors:* Booyabazooka, Angela, Tsavage, Gardenmaster, Toytoy, Lulu of the Lotus-Eaters, Xezbeth, Vortexrealm, Paleorthid, Tonners62, Cryptic, Mikeblas, Pegship, Arthur Rubin, SmackBot, Gobonobo, Amalas, Nigholith, Malljaja, Aphilo, Fadesga, Surtsicna, Lightbot, Yobot, Wotnow, Frietjes, Helpful Pixie Bot, CitationCleanerBot and Anonymous: 4

2, Andrea105, CalicoCatLover, Mr. Anon515, John of Reading, WikitanvirBot, Putrefaction, GoingBatty, Tommy2010, Asephei, Jesanj, L Kensington, Inka 888, Ariel1024, Entomology Lab, Hbreton19, JenCom, 28bot, Whoop whoop pull up, ClueBot NG, Cwmhiraeth, Jack Greenmaven, Gilderien, Satellizer, Cj005257, Dictabeard, Estopedist1, Widr, Helpful Pixie Bot, In actu, BG19bot, Dylanarmonchasepassword, MusikAnimal, AwamerT, Mark Arsten, Glacialfox, Achowat, Aisteco, Mark Laguitan, Mewuzhere, Capthook11, SaudiPseudonym, Pratyya Ghosh, Ddcm8991, Timelezz, Lugia2453, Frosty, SFK2, Kmanuola, 069952497a, Evermist509, Jewnicorn592, Mykophile, Justacheck, Meteor sandwich yum, Lillers101, CatcherStorm, Monkbot, JurrasicPark67, Ingleburnhs, Brainblob107, Emorillas and Anonymous: 477

- **Dillo Dirt** *Source:* https://en.wikipedia.org/wiki/Dillo_Dirt?oldid=669252981 *Contributors:* Ground Zero, TexasAndroid, Weasel5i2, XLinkBot, Lightbot, Yobot, Jonathan Haas, Shadowjams, FrescoBot, Metricmike and Anonymous: 7

- **Ecuador composting method** *Source:* https://en.wikipedia.org/wiki/Ecuador_composting_method?oldid=644640378 *Contributors:* Smack-Bot, Juandev, Vrac, Addbot and Jb0007

- **Eisenia fetida** *Source:* https://en.wikipedia.org/wiki/Eisenia_fetida?oldid=638334628 *Contributors:* Quercusrobur, Lumos3, Jredmond, Worm-Runner, Lesgles, Discospinster, Guettarda, 99of9, Zachlipton, Jgbundy, Stemonitis, Barrylb, The Lightning Stalker, T34, Tar-Palantir, Bremen, Eubot, Gdrbot, Pseudomonas, GeeJo, JHCaufield, Caerwine, Carabinieri, SmackBot, Stubblyhead, Uthbrian, Abrahami, Elapied, Kaarel, Courcelles, Difluoroethene, Bposert, DumbBOT, Bheal, JamesAM, Thijs!bot, Neil916, Nick Number, JamesBWatson, Bob, Jeff G., Soliloquial, TXiKiBoT, Raymondwinn, Mooreds, Logan, Flyer22, Jojalozzo, OKBot, Fullobeans, Wanderer57, Ziromar222, DragonBot, BOTarate, Lokionly, XLinkBot, Facts707, ErkinBatu, Addbot, Gvfarns, Flakinho, Zorrobot, Luckas-bot, Yobot, IRP, Galoubet, Materialscientist, Zad68, Craig Pemberton, Stephen Morley, Pinethicket, Robertsjk, Wikielwikingo, Dinamik-bot, Sideways713, Mean as custard, RjwilmsiBot, WikitanvirBot, ZéroBot, Medeis, Vaibhavinfo, ClueBot NG, Hockeyman1997, Calabe1992, Salma Vian, Sorryinputerror, Mjfannnnnn, Federales, Howunusual and Anonymous: 66

- **Fairfield Materials Management Ltd** *Source:* https://en.wikipedia.org/wiki/Fairfield_Materials_Management_Ltd?oldid=564033079 *Contributors:* Rich Farmbrough, DGG, Robert Skyhawk, Tnxman307, Mhockey, Addbot, Yobot, Krysjw, DrilBot, Hmainsbot1 and Anonymous: 2

- **Grasscycling** *Source:* https://en.wikipedia.org/wiki/Grasscycling?oldid=656913436 *Contributors:* Alan Liefting, Elf, Rjwilmsi, Stephenb, Shawnc, Vranak, Zodon, Westexaslawnman, Yobot, Jimbob16314, EmausBot, Look2See1, H3llBot, Mohamed-Ahmed-FG, Green Turf and Anonymous: 3

- **Hermetia illucens** *Source:* https://en.wikipedia.org/wiki/Hermetia_illucens?oldid=685292306 *Contributors:* Wetman, Chowbok, EugeneZelenko, Stemonitis, BD2412, Eubot, Gdrbot, Triku~enwiki, The Rambling Man, Brandmeister (old), Grafen, SmackBot, Notafly, Wikiklaas, Agradman, Stst399, Dicklyon, MGlosenger, Teratornis, Jaerik, Dyanega, Thijs!bot, Leuko, Magioladitis, Dp76764, JaGa, CliffC, TyrS, Cyanolinguophile, Qatter, Maproom, Rolivier79, Muhammad Mahdi Karim, Hiyowassup, Bardobro, WereSpielChequers, Oculi, Heds, Trfasulo, Noca2plus, XLinkBot, Lumenos, Imagine Reason, Nepenthes, Addbot, DOI bot, Dawynn, Leszek Jańczuk, MrOllie, Yobot, Julia W, Yngvadottir, AnomieBOT, Götz, Gee W., EHRice, RobertEves92, SailorOnHorseback, Xqbot, TechBot, Control.valve, Anmorata, FrescoBot, Simuliid, Archaeodontosaurus, Citation bot 1, Gingin Boy, 777sms, Vermisapiens, EmausBot, John of Reading, WikitanvirBot, Marinos1977, Vetrider, ClueBot NG, Sharktopus, Cwmhiraeth, Animalspecialties, Eef12, Sabercrombie, Nishaca, Helpful Pixie Bot, Slanders13, Ericsheldon, KLBot2, Editor Bert, Mgrogger, NotWith, Thunder42strike, Rudork, Liam987, Maxasher, Stratiomyidae, Tortie tude, Smartedits5, EvMsmile and Anonymous: 49

- **Hotbed** *Source:* https://en.wikipedia.org/wiki/Hotbed?oldid=607476586 *Contributors:* Anthony Appleyard, BDD, Woohookitty, Synthebot, Atubeileh, Three-quarter-ten, Addbot, Xqbot, DennisIsMe, CocuBot, Clone200 and Anonymous: 2

- **Humic acid** *Source:* https://en.wikipedia.org/wiki/Humic_acid?oldid=684270182 *Contributors:* Malcolm Farmer, Lexor, Ike9898, Eequor, Onco p53, Ary29, DanielCD, Cacycle, Zombiejesus, Vsmith, Vortexrealm, DaveGorman, Paleorthid, Velella, Dave.Dunford, BlueCanoe, Alvis, RHaworth, Dj Capricorn, Debivort, YurikBot, Scott Teresi, Pburka, Stephenb, Frenkmelk, Dspradau, Garion96, SmackBot, Jrockley, Bluebot, George Church, Deli nk, Uthbrian, A. B., Fuhghettaboutit, Drphilharmonic, Xiutwel, Jaganath, Beetstra, TastyPoutine, Lesfreck, Tawkerbot2, Chrumps, Harej bot, Nick Wilson, Gogo Dodo, Alaibot, Thijs!bot, Caseyc, TimVickers, Mikenorton, Brewhaha@edmc.net, Nono64, Rod57, STBotD, Pdcook, Buyoaks1, Lamro, Petergans, SieBot, Tresiden, ترجمان05, Mattmnelson, Chem-awb, ClueBot, Intelitem~enwiki, Mild Bill Hiccup, Tadhussey, Capewellmj, Coinmanj, UrsoBR, SchreiberBike, XLinkBot, NellieBly, Prbloom, Addbot, Fainzweig, LatitudeBot, Ronhjones, CanadianLinuxUser, Debresser, Da best editor, Ben Ben, Luckas-bot, Yobot, Apophenic, AnomieBOT, Götz, Materialscientist, Citation bot, TheAMmollusc, Dr Fulvic, Jeffwend, Citation bot 1, Trappist the monk, EmausBot, Gzuufy, Fixertx, Rocketrod1960, ClueBot NG, มือใหม่, CaptionOrganic, Helpful Pixie Bot, Ralf Ostertag, Ffejmopp, Halfb1t, Aggregatation, ChrisGualtieri, Zorahia, Capt.lingard, Shilajit Pure Resin and Anonymous: 88

- **Humus** *Source:* https://en.wikipedia.org/wiki/Humus?oldid=663924539 *Contributors:* Marj Tiefert, Sodium, Tarquin, Malcolm Farmer, --April, Jpsturm, Anthere, Perique des Palottes, Quercusrobur, Gdarin, Wapcaplet, Looxix~enwiki, Александър, Glenn, Nikai, Nohat, 1984, WormRunner, Sverdrup, Academic Challenger, Antonin~enwiki, Arm, Pengo, Alan Liefting, DocWatson42, Fennec, MathKnight, Varlaam, Joshuapaquin, Jrdioko, Utcursch, Sam Hocevar, D6, Rich Farmbrough, Eric Shalov, Kwamikagami, Vortexrealm, Microtony, Jumbuck, Alansohn, Ryanmcdaniel, 119, Improv, Paleorthid, Daniel.inform, Bart133, Kazvorpal, Nuno Tavares, Pekinensis, Consequencefree, Robert K S, WadeSimMiser, JeremyA, SCEhardt, Mandarax, Graham87, Rjwilmsi, Jake Wartenberg, Feydey, Daniel Collins, Kalogeropoulos, DoubleBlue, FlaBot, Musical Linguist, RexNL, WriterHound, YurikBot, Peter G Werner, Anomalocaris, TDogg310, CLW, 21655, JoanneB, Höyhens, SmackBot, Telestylo, Hardyplants, Gilliam, Squiddy, Jerome Charles Potts, MichaelBillington, IrisKawling, Drphilharmonic, DavidHallett, Yeomansplowchris1, Lambiam, Dave314159, Mgiganteus1, Brazucs, KokomoNYC, Fluppy, Casull, Courcelles, Tawkerbot2, MyUsername, JForget, JohnCD, Kazubon~enwiki, Christian75, Oliver202, Marek69, John254, Mmcknight4, Escarbot, RapidR, Smartse, Brucelaidlaw, Spencer, Gökhan, Barek, Smulthaup, Edwardspec TalkBot, Leolaursen, Magioladitis, Freedomlinux, VoABot II, LafinJack, Mtiffany71, DerHexer, JaGa, Nono64, LedgendGamer, J.delanoy, Mike.lifeguard, Katalaveno, Janus Shadowsong, Coin945, STBotD, Bonadea, Inwind, VolkovBot, Abbail, Steven J. Anderson, DennyColt, ^demonBot2, Madhero88, Red58bill, PGWG, SieBot, Gerakibot, TOMolloy, Keilana, Faradayplank, Soyseñorsnibbles, ClueBot, Alandmanson, DragonBot, DOHill, Manzzzz, Elizium23, Versus22, XLinkBot, Addbot, Download, Glane23, AndersBot, AgadaUrbanit, Numbo3-bot, Jfponge, Theking17825, MZaplotnik, Anxietycello, OlEnglish, Bartledan, Luckas-bot, Yobot, II MusLiM HyBRiD II, AnomieBOT, 1exec1, Killiondude, Flewis, Materialscientist, Citation bot, Ilikecupcakes42, Neurolysis, ArthurBot, Pkravchenko, Arkadia96, RibotBOT, Mathonius, FrescoBot, Tangent747, Kyle194856874, Pinethicket, I dream of horses, HRoestBot,

RedBot, FoxBot, Oursspace, DARTH SIDIOUS 2, Mean as custard, EmausBot, Look2See1, K6ka, Auró, ZéroBot, Josve05a, Micro2, Aeonx, Tolly4bolly, Erianna, ChuispastonBot, Berberisb, ClueBot NG, Gilderien, Doh5678, Brickmack, Widr, Helpful Pixie Bot, Hotpop2, Guará-wolf, Stelpa, Ricordisamoa, Zedshort, Klilidiplomus, ChrisGualtieri, Hridith Sudev Nambiar, AutomaticStrikeout, Mantroller, Epicgenius, Jianhui67, TerryAlex, Guanalan100, AnimalGoToGirl cx, Mikey9363 and Anonymous: 204

- **John Innes compost** *Source:* https://en.wikipedia.org/wiki/John_Innes_compost?oldid=643510805 *Contributors:* William Avery, Bearcat, GraemeLeggett, Malcolma, Kingboyk, Jonobennett, Jamse, Iridescent, IceDragon64, Red58bill, SieBot, Niceguyedc, Addbot, MrOllie, Yobot, Sickymicky, Fairwyn, Cmwilson7 and Anonymous: 1

- **Keyhole garden** *Source:* https://en.wikipedia.org/wiki/Keyhole_garden?oldid=676461427 *Contributors:* WAS 4.250, EmausBot and Proglin

- **Leaf mold** *Source:* https://en.wikipedia.org/wiki/Leaf_mold?oldid=632671313 *Contributors:* Bryan Derksen, Stephen Gilbert, Rgamble, Quercusrobur, Infrogmation, Stan Shebs, Glenn, Jose Ramos, Sheridan, Guettarda, Enric Naval, Vortexrealm, Paleorthid, Jwanders, Wavelength, Splash, Welsh, BorgQueen, SmackBot, Pgk, Cacuija, Shai-kun, Squiddy, JohnCub, Lucy Cassidy, Croton, Soulbot, J A Ellam, Jeepday, Fences and windows, Red58bill, SieBot, FerdinandFrog, Lokionly, Addbot, Lightbot, Yobot, Apothecia, Anna Frodesiak, Wikininja2.0, Finalius, Look2See1 and Anonymous: 18

- **Mulch** *Source:* https://en.wikipedia.org/wiki/Mulch?oldid=669360846 *Contributors:* Rmhermen, Quercusrobur, Kchishol1970, Jizzbug, Ronz, Angela, Taxman, TMLutas, Tsavage, Pengo, DocWatson42, Mboverload, Jrdioko, Pgan002, Onco p53, Vic Fontaine, Mcpusc, Smalljim, Vortexrealm, Richi, Alansohn, Malo, Velella, Tony Sidaway, Rzelnik, BlueCanoe, Firsfron, WadeSimMiser, Pdn~enwiki, BD2412, Rjwilmsi, Vegaswikian, Soredewa, Hlodynn, Wavelength, Pip2andahalf, Crazytales, Splash, RadioKirk, Leighblackall, Ozzykhan, Wiki alf, Zwobot, CDA, Jtc, Tevildo, Luiscolorado, David Biddulph, Rwellington, ChemGardener, SmackBot, Bigbluefish, ASarnat, Hardyplants, Isaac Dupree, Bluebot, Jeekc, Yorick8080, EdGl, DMacks, SilkTork, Tdudkowski, --colibri--, Phasmatisnox, CmdrObot, CWY2190, DumbBOT, Mattisse, Thijs!bot, Epbr123, Pstanton, Mojo Hand, Stybn, Mulch-guide, Justgardens, Seaphoto, QuiteUnusual, CZmarlin, Hopiakuta, Figma, Steveprutz, Steven Walling, OldPine, EagleFan, U608854, JaGa, Flami72, Pere prlpz, Gandydancer, Brspenn, NAHID, R'n'B, Nono64, Transisto, Extransit, Jeepday, Naniwako, Danger Pain, Pamon, DorganBot, Bocilla, Sam Blacketer, BigStickCharly, VolkovBot, D36csr, ReadersFavorite, TXiKiBoT, Frankenphile, RobertPlamondon, AlleborgoBot, Red58bill, Bobob1916, Vanished User 8a9b4725f8376, Chrisc2168, Lightmouse, Staylor3, Shooter16101, Wahrmund, ClueBot, GorillaWarfare, Petervanzelst, MATThematical, Hysocc, Drmies, Sparrowhawkone1, Taranet, Dream belive, XLinkBot, Lumenos, Dthomsen8, Addbot, CanadianLinuxUser, MrOllie, Lightbot, Krano, Luckas-bot, JackPotte, Specious, AnomieBOT, AdjustShift, Citation bot, ArthurBot, Bristles12, Ahmadrknowledge, Capricorn42, Drilnoth, Anna Frodesiak, Gari5737, JadeInOz, FrescoBot, Bfchuck, Greenbrow, SouthernMNrabbitrescue, Stephen Morley, Pinethicket, Modeltcentral, RedBot, Pikiwyn, DixonD-Bot, Dinamik-bot, DARTH SIDIOUS 2, NameIsRon, EmausBot, WikitanvirBot, Rasputin72, Logical Cowboy, Look2See1, ZéroBot, Arpandey, ClueBot NG, PaleCloudedWhite, Helpful Pixie Bot, TopDog5450, Rowan Adams, Mulching, Darorcilmir, Epicgenius, Henri T J, Landscapefan1251, Padraig Singal, Monkbot, Jbauer99, MEJones1, Koios Titan of Winter, Tessaltvater, Angusrobertcheyne and Anonymous: 164

- **Multrum** *Source:* https://en.wikipedia.org/wiki/Multrum?oldid=506723292 *Contributors:* Bearcat, Carllindstrom, Katharineamy, Bonadea, Deutschgirl, JamietwBot and ChrisGualtieri

- **Nematode** *Source:* https://en.wikipedia.org/wiki/Nematode?oldid=684518514 *Contributors:* AxelBoldt, Magnus Manske, Lee Daniel Crocker, Mav, Josh Grosse, PierreAbbat, Azhyd, Hephaestos, Michael Hardy, Lexor, Kku, Ixfd64, Zanimum, Pcb21, Ahoerstemeier, Julesd, Emperorbma, Timwi, Janko, Greenrd, Tpbradbury, Saltine, Nv8200pa, PuzzletChung, Robbot, Hankwang, Nyh, Academic Challenger, Rorro, SaraS~enwiki, Hadal, Wikibot, GerardM, Dmn, Jholman, Marc Venot, DocWatson42, Everyking, Wmahan, Gdr, Mr d logan, Antandrus, Williamb, Kaldari, Sean Heron, DNewhall, Phil1988, Sam Hocevar, Gscshoyru, Joyous!, Alperen, Adashiel, Mike Rosoft, Discospinster, Rich Farmbrough, TedPavlic, JoeSmack, Violetriga, Eric Forste, Ascorbic, Kwamikagami, Mauler, CDN99, Bobo192, Dragon76, Smalljim, Shenme, Vortexrealm, Arcadian, Deryck Chan, Googie man, Alansohn, Guy Harris, Quatermass, MauriceReeves, Riana, Seans Potato Business, AtonX, House of Shin, Bart133, Radical Mallard, Nathanlarson32767, RainbowOfLight, Shoefly, Kazvorpal, Stuartyeates, Stemonitis, Borrel, Pekinensis, Woohookitty, Qaddosh, WadeSimMiser, GregorB, Karmosin, Cyberman, Dysepsion, Mandarax, RichardWeiss, Graham87, BD2412, Canderson7, Rjwilmsi, Koavf, Harry491, Quiddity, Kazrak, Ligulem, Yamamoto Ichiro, FlaBot, SchuminWeb, Eubot, LiquidGhoul, RexNL, Gurch, Alphachimp, McDogm, Chobot, Michaelritchie200, DVdm, Gdrbot, Bgwhite, Dj Capricorn, WriterHound, Gwernol, YurikBot, RobotE, Neitherday, Pigman, Stephenb, CambridgeBayWeather, Anomalocaris, MosheA, NawlinWiki, Dysmorodrepanis~enwiki, Wiki alf, Bou, Howcheng, Apokryltaros, Nick, Daniel Mietchen, Bob0theOmighty, E rulez, Mooncowboy, Zwobot, Supspirit, Trainra, CLW, Tigershrike, WAS 4.250, Joshua368, Open2universe, Lt-wiki-bot, Nikkimaria, Ketsuekigata, Dynamaniac, Junglecat, IanRiley, Zvika, CIreland, SmackBot, Snickersnee, KHenriksson, Bomac, Davewild, Clpo13, RedSpruce, Jrockley, Floydspinky71, Canthusus, Gaff, Gilliam, Skizzik, Eug, Chris the speller, Persian Poet Gal, Miquonranger03, Hibernian, Darth Panda, Rlevse, Mike hayes, Can't sleep, clown will eat me, Jeffire, Abyssal, Addshore, Bardsandwarriors, Grover cleveland, Memming, Aldaron, Tvaughn05, Eleiser, TheLimbicOne, EStoner, Dreadstar, RandomP, Jóna Þórunn, J.smith, SashatoBot, Nishkid64, Valfontis, Soap, Nights Not End, AmiDaniel, Marco polo, Gobonobo, Epingchris, Kevmin, Joelmills, Butko, Mgiganteus1, Smith609, Waggers, Dalstadt, Funnybunny, RMHED, Cerealkiller13, Sifaka, BranStark, Nonexistant User, Iridescent, JMK, Myopic Bookworm, Paul venter, Wfgiuliano, Kaarel, Joseph Solis in Australia, Tmangray, Newone, Takarada, Kalessin11, Courcelles, Tawkerbot2, Talono, IronChris, Danberbro, Tanthalas39, Jom~enwiki, Dgw, AshLin, MrFish, Namayan, AndrewHowse, Cydebot, Generalnonsensecomic, Kanags, Ryan, Nick Wilson, Anthonyhcole, Chasingsol, Tawkerbot4, Christian75, Asenine, Narayanese, Void main, Hbah427, Vanished User jdksfajlasd, Rbanzai, Casliber, JamesAM, Thijs!bot, Epbr123, NewInn, Momogonz, Feud, Headbomb, Luigifan, Marek69, John254, Sturm55, OrenBochman, LachlanA, Mentifisto, Hmrox, AntiVandalBot, Quintote, Stug.stug, Smartse, Danger, MikeLynch, Deflective, Husond, MER-C, Instinct, OhanaUnited, Andonic, Igodard, Kerotan, Acroterion, WolfmanSF, VoABot II, JNW, Yandman, Michael Goodyear, Soulbot, BrianGV, Allstarecho, Japo, Styrofoam1994, Hveziris, Rolf Schmidt, Hintswen, S3000, Darth.maul.is.alive, MartinBot, Bissinger, Anaxial, Zany zacky, Tgeairn, Artaxiad, Slugger, J.delanoy, CFCF, Trusilver, Philcha, Adavidb, Boghog, Dbiel, Extransit, Rufous-crowned Sparrow, Shawn in Montreal, Katalaveno, Enuja, Jeepday, Ephebi, Mattximus, AntiSpamBot, Plasticup, Chiswick Chap, Twalden4, NewEnglandYankee, Dividing, ARTE, Hanacy, Sarregouset, Xaxx, Gmahan, Psamathos, Meiskam, VolkovBot, BabySinclair, DrMicro, AlnoktaBOT, Vlmastra, TXiKiBoT, Agrihouse, Rei-bot, Qxz, Someguy1221, Browneee, Clarince63, ^demonBot2, Aliasxerog, Thedogroxie, Dirkbb, Avolovisky, Matthewautocraze, Dhorspool, Monty845, Twooars, Doc James, Nagy, Stacy734, Red58bill, Sten for the win, Ruraloccur, Brennoncds11, SieBot, Coffee, Tiddly Tom, Curlyred, Cyberix, Jharris0221, Redhookesb, Keilana, Myayize, Sksharma1972, FunkMonk, Flyer22, Radon210, JudgeSpear, Blaireaux, Yerpo, Bluenema, Massacre99, Oxymoron83, PhilMacD, Alex.muller, AlanUS, Soulweaver, Mrthescholz, Imac.vincent, Mr. Stradivarius, Joshschr, Bpeps, Denisarona, ClueBot, Trfasulo, Weisbah, The

Thing That Should Not Be, Matdrodes, Punyasloke, Jan1nad, Drmies, Firth m, Excirial, Robbie098, XCalPab, Coralmizu, Monobi, Adimovk5, Shinkolobwe, JamieS93, Mikaey, SchreiberBike, Πβ~enwiki, Thingg, Been001, Aitias, 7, Footballfan190, Subash.chandran007, Versus22, Katanada, Vanished User 1004, ZephyrGreene, Moolowdy76, Facts707, Skarebo, WikHead, Badgernet, Alexius08, Noctibus, LeDiableBrun, Amynta, Thatguyflint, Addbot, C6541, Some jerk on the Internet, DOI bot, Fyrael, Landon1980, Hunt91, Ronhjones, CanadianLinuxUser, Leszek Jańczuk, Download, Glane23, Favonian, Kyle1278, Peti610botH, Lgg21, Tide rolls, Jan eissfeldt, Romanskolduns, Gail, Quasibr, Porcellio, Luckas-bot, Yobot, Jason Recliner, Esq., Guarana9, Guy1890, Parnacio, Marshall Williams2, Retro00064, AnomieBOT, KDS4444, Juzhong, Fatal!ty, Apollo1758, JackieBot, Teratyke, Kingpin13, Пика Пика, Materialscientist, Cecole, OllieFury, Frankenpuppy, LilHelpa, Obersachsebot, Xqbot, Bihco, Gigemag76, Renaissancee, Anna Frodesiak, Abce2, SassoBot, Shadowjams, Miyagawa, Samwb123, Yeahmean97, Dougofborg, Thehelpfulbot, Nenya17, BoomerAB, FrescoBot, Peyton415, Mshearn, Jc3s5h, Flomawo, D'ohBot, VI, Pikachuwashere, Mrseibert, Drew R. Smith, Louperibot, Citation bot 1, NoNamer123, Slobodan Grasic, Pinethicket, PrincessofLlyr, Jonesey95, Nightsmaiden, Serols, Florescent, Yakovlev Yegor, Kibi78704, Reconsider the static, Tim1357, Animalparty, Spikanorx3, Dinamik-bot, Obsidian Soul, Mean as custard, RjwilmsiBot, Bento00, B-Boy 369, EmausBot, WikitanvirBot, Immunize, Look2See1, Pdeley, Super48paul, Yt95, Tragicspade, Tommy2010, HiW-Bot, ZéroBot, John Cline, Josve05a, Akerans, Pathay, Maddendalybrokaw, Ebrambot, Fionaisawesome, Ajaxdslayer, Ocean Shores, Augurar, Donner60, Gongoozler123, ChuispastonBot, DASHBotAV, JonRichfield, ClueBot NG, Ykvach, Jack Greenmaven, Jasonf1980, Gilderien, Bped1985, PaleCloudedWhite, Anvi009, Guillaume8507, Frietjes, Muon, Mesoderm, Rezabot, Hungry Vampire, Jorgenev, DBigXray, BG19bot, Vegas Viper, 959worm, NotWith, Earendil56, Zedshort, Gug01, Lobbygow, BattyBot, Jeanloujustine, ChrisGualtieri, Ducknish, Dexbot, Br'er Rabbit, Sydneyrocks247, Numbermaniac, Frosty, Jamesx12345, Ilovemoscow, Qwertyuiiopasdfghjkl, The Anonymouse, Jerkwad25, Sharon santhosh123, SoWhAt249, Blsmith21, Blythwood, Monkeypolice000, Thenemes, MSperbeck, Jwratner1, Snbarnes, Coreyemotela, Vpandrangi, JaconaFrere, G S Palmer, Lesbowen, Phleg1, Monkbot, Ubershweet628, 空间的拓荒者, Dpwiese, Pwiesethetruth, Ashleyearley, EvMsmile, Weegeerunner, Clicksm, KasparBot, Annamyus user, Whmbkx and Anonymous: 734

- **Night soil** *Source:* https://en.wikipedia.org/wiki/Night_soil?oldid=683641579 *Contributors:* Edward, Palfrey, Jogloran, Furrykef, Altenmann, KellyCoinGuy, Andycjp, Geni, Antandrus, Eisnel, Cfailde, ESkog, Fenevad, Art LaPella, Vortexrealm, La goutte de pluie, Kjkolb, Jemfinch, Anthony Appleyard, Guy Harris, Paleorthid, Hunter1084, KingTT, Saga City, Squidley, Graham87, Miserlou, Diogenes00, Jimp, DMahalko, Bachrach44, Malaiya, Hunnyhiteshseth, Jonathan.s.kt, SmackBot, Bluebot, Thumperward, Frecklegirl, Liontooth, Al95521, Pwjb, Sljaxon, Parrot of Doom, Fremte, Nygdan, Gregorydavid, Meco, Shawn D., Quodfui, TeflonSoul, Cyhawk, Altfish, Malleus Fatuorum, Amity150, Ericjs, PhiLiP, Andonic, AtticusX, GuelphGryphon98, GTZ-44-ecosan, Lamaybe, Gurchzilla, Juliancolton, 2eet2eet, Riffraffselbow, Red58bill, Kapalama, Bielle, Patrick Nevin, Kinkyturnip, ClueBot, Richerman, Trivialist, Gtstricky, Millionsandbillions, Churchill317, Lumenos, Mitch Ames, Clubheard~enwiki, Ewbie, Hoplophile, Addbot, Backblow, MrOllie, Lightbot, Jarble, Yobot, Donfbreed, AnomieBOT, The sock that should not be, Jmundo, Douchedoom, HRoestBot, Serols, Chromatikoma, Lotje, Consultant09, Kangesh, Ra1n, BrendanFrye, GoingBatty, Ericblair109, Doorautomatica, MajorVariola, L Kensington, Hairyleafman, Hasansara3, George Ponderevo, Lawfriedrich, Lizwe sithole, Captain Cornwall, Beanstash, EvMsmile, JMWt and Anonymous: 86

- **Oligochaeta** *Source:* https://en.wikipedia.org/wiki/Oligochaeta?oldid=680376997 *Contributors:* Josh Grosse, Dan Koehl, Muriel Gottrop~enwiki, Bogdangiusca, Hike395, Grendelkhan, Amphioxys, WormRunner, Naddy, UtherSRG, Abigail-II, Pgan002, JoJan, MakeRocketGoNow, Discospinster, Rich Farmbrough, Bender235, Kwamikagami, Bobo192, Vortexrealm, ChriKo, Scentoni, Cronus, Wtmitchell, Adrian.benko, Stemonitis, Munificent, T34, Magister Mathematicae, Rjwilmsi, Koavf, Harry491, Eubot, TeaDrinker, Gdrbot, Dj Capricorn, YurikBot, Hairy Dude, Jimp, RussBot, Dysmorodrepanis~enwiki, Xabian40409, Asterion, SmackBot, Bomac, Tarkya, TheLimbicOne, MTSbot~enwiki, Bruinfan12, Timichal, Thijs!bot, Neil916, Nick Number, Shirt58, JAnDbot, Deflective, MasterA113, .anacondabot, Bennybp, Silentaria, The cattr, DerHexer, Hbent, Patstuart, MiltonT, Anaxial, Justin, Barrie gm jamieson, Gurchzilla, Winderful1, Andersman, Zomitra, VolkovBot, CWii, AlnoktaBOT, Philip Trueman, BotKung, GoTeamVenture, IDNeon, Ark2120, Alexbot, Estirabot, SchreiberBike, Esoxid, ZooFari, Addbot, SpBot, Notpookie, Zorrobot, Yobot, Coffinfly, KDS4444, Materialscientist, Citation bot, Xqbot, TheAMmollusc, Jlaverdure, Miracle Pen, RjwilmsiBot, TjBot, EmausBot, Maxim Gavrilyuk, ZéroBot, ChuispastonBot, ClueBot NG, Movses-bot, Widr, Helpful Pixie Bot, AvocatoBot, SFK2, EszterJK, Ritzner von Dracul, Comaniciu David and Anonymous: 65

- **Olive mill pomace** *Source:* https://en.wikipedia.org/wiki/Olive_mill_pomace?oldid=649633941 *Contributors:* KillerChihuahua, RHaworth, Pdcook, Zefr, Voxii, Japsmp, Look2See1, Julietdeltalima and Anonymous: 1

- **Organopónicos** *Source:* https://en.wikipedia.org/wiki/Organop%C3%B3nicos?oldid=672750519 *Contributors:* Edward, GTBacchus, Jengod, Owen, Alan Liefting, Pgan002, Forbsey, Bobrayner, Rjwilmsi, Lockley, Jrtayloriv, Zotel, Takethemud, MichaelW, SmackBot, Bluebot, Wirtheim, Zleitzen, SilkTork, Robofish, Ayanoa, Nunquam Dormio, W.0q, Drm310, Dmcamp, Universaladdress, Lucasbfrbot, Transcona Slim, FusionNow, Gardenparty, Addbot, Kilom691, AnomieBOT, LucienBOT, RjwilmsiBot, TjBot, Wikipelli, ClueBot NG, MickStep, BG19bot, Sidelight12, VanishedUser000000000 and Anonymous: 19

- **Sebakh** *Source:* https://en.wikipedia.org/wiki/Sebakh?oldid=604508132 *Contributors:* DopefishJustin, Pjamescowie, CatherineMunro, Viajero, Altenmann, Blainster, Canadabear, PDH, Adamsan, CanisRufus, Kwamikagami, Cohesion, GUS JOHN GEORGE, Splash, DanMS, Sloman, Xmts, Colonies Chris, Addbot, Vyom25, HoremWeb, VernoWhitney, Y-barton, Toffanin, Bonios, APerson and Anonymous: 11

- **Spent mushroom compost** *Source:* https://en.wikipedia.org/wiki/Spent_mushroom_compost?oldid=661924089 *Contributors:* Quercusrobur, Jose Ramos, Alan Liefting, Nunh-huh, PDH, Guanabot, Vortexrealm, Stemonitis, Jwanders, Dwaipayanc, Wavelength, Splash, Dialectric, SmackBot, Squiddy, Rkmlai, Brynnar~enwiki, PhilKnight, Qatter, Red58bill, Addbot, ThinkingTwice, AnomieBOT, Citation bot, Anna Frodesiak, Look2See1, Blonder1984, Helpful Pixie Bot, KLBot2 and Anonymous: 6

- **Stubble-mulching** *Source:* https://en.wikipedia.org/wiki/Stubble-mulching?oldid=421053266 *Contributors:* Trevor MacInnis, Agradman, Jstreutker, Dthomsen8, MatthewVanitas and Erik9bot

- **Used coffee grounds** *Source:* https://en.wikipedia.org/wiki/Used_coffee_grounds?oldid=675759252 *Contributors:* Bearcat, Magioladitis, JamesBWatson, Bryce Carmony, NotWith, Bananasoldier and FreddieBarlett

- **Uses of compost** *Source:* https://en.wikipedia.org/wiki/Uses_of_compost?oldid=640388870 *Contributors:* Stone, Alan Liefting, SECProto, Woohookitty, Rjwilmsi, Salix alba, RussBot, Phorque, Avalon, That Guy, From That Show!, SmackBot, Elonka, Chris the speller, LLucas, MER-C, Repku, Meredyth, Lamro, Red58bill, Gardenmandy, Crowsnest, Karanne, Nutriveg, Capricorn42, J04n, Jorden12, Look2See1, Wayne Slam, Hotsunroses, Braincricket, IvoryMeerkat, Corteli9, EvMsmile and Anonymous: 9

- **Vermicompost** *Source:* https://en.wikipedia.org/wiki/Vermicompost?oldid=678581760 *Contributors:* Bryan Derksen, Tarquin, Alex.tan, Chuckhoffmann, Fcueto, William Avery, Quercusrobur, (, Ronz, Jeandré du Toit, Jose Ramos, Pollinator, WormRunner, Rhombus, Pengo, Alan Liefting, Pne, Pgan002, Beland, Madmagic, OverlordQ, PDH, Neschek, Rich Farmbrough, Cacycle, Vsmith, Gjm, CanisRufus, Greenmoss, Talbrech, Cmdrjameson, Petronivs, Mrdude, Vortexrealm, Maurreen, Macho, Wtmitchell, Gene Nygaard, Bookandcoffee, ChrisJMoor, Mindmatrix, Barrylb, Davidkazuhiro, Jwanders, Bikeable, Icey, Rjwilmsi, SeanMack, Parutakupiu, WriterHound, Wavelength, Sillybilly, Bhny, Splash, Epolk, SpuriousQ, Pseudomonas, Alynna Kasmira, NawlinWiki, Jonathan Webley, Jbreazeale, Stefeyboy, TCH, Ke6jjj, Allens, DasBub, ChemGardener, SmackBot, KVDP, Zekkelley, Ohnoitsjamie, Andy M. Wang, Tv316, Chris the speller, SB Johnny, Carbonrodney, Deli nk, A. B., Brimba, G716, DMacks, L337p4wn, Gobonobo, Cmh, FrostyBytes, Rkmlai, Citicat, Peter Horn, RMHED, HisSpaceResearch, Iridescent, Ayanoa, DBooth, Christian025, Tex, Odie5533, Psuliin, Outdoorvegan, Shirulashem, Bposert, DumbBOT, Thijs!bot, Fisherjs, Pjvpjv, Neil916, Grayshi, Stannered, Chubbles, Sofia Roberts, KP Botany, Jj137, Danger, Kleomarlo, Stevenger, Adjwilley, Greg Comlish, PhilKnight, JamesBWatson, Animum, Sgr927, ArchStanton69, WLU, Wassupwestcoast, Rettetast, Boston, Idontthinkso, JuniperFuse, Stefandrew, Notreallydavid, Skier Dude, Leia tyndall, AntiSpamBot, (jarbarf), Belovedfreak, Jorfer, Jpmacke, Mleonard85032, Inwind, Idiomabot, Remi0o, Sam Blacketer, Christophenstein, Wikibmike, TXiKiBoT, Rei-bot, ElinorD, Ychastnik APL, Vijay68, Cremepuff222, Mooreds, Red58bill, Christidrick, SieBot, Tiddly Tom, Caulde, WereSpielChequers, Jauerback, Winchelsea, Victorcoutin, Dburdenbates, Industrialg33k, Flyer22, Bob98133, Dillard421, Anakin101, FredrikLähnn, WormPower, ClueBot, Mild Bill Hiccup, Shabbychef, Artemis enaid, Csrwizard, Muditpurohit, Micha, Mightycord, Versus22, Vanished user uih38riiw4hjlsd, Australorp, XLinkBot, Steeljack, TFOWR, CalumH93, Addbot, MrOllie, CarsracBot, Arben1951, Tassedethe, OlEnglish, Tempodivalse, Backslash Forwardslash, AnomieBOT, Glennybee, Leangreenhome, L3lackEyedAngels, Materialscientist, Puddypie007, Wog400, Xqbot, Apothecia, Pnuematics, Brynnab, Fernyburn, Twirligig, Zdgnet, Newnorb, BoomerAB, Stonelakefarm, Wormwrangler, Karincaprice, Bhoeschcod, PigFlu Oink, Amshdoc, Sandcat01, Pinethicket, 19saraswathi05, Robertsjk, Σ, Pinochet (3), Kirstendirksen, Animalparty, Clarkcj12, Blueacre, Minimac, Vermisapiens, Fiftytwo thirty, Happydranch1, EmausBot, Wikipelli, Josve05a, AManWithNoPlan, Kindzmarauli, Jyothiprakashhk, 28bot, ClueBot NG, O.Koslowski, خردمندان, Newyork1501, Gob Lofa, Gauravjuvekar, Northamerica1000, Hippiehiker5091, Cclehnen, Cyberbot II, Khazar2, Sminthopsis84, Bananasoldier, PhantomTech, Cloudyjbg27512, Lagoset, Monkbot and Anonymous: 234

- **Windrow composting** *Source:* https://en.wikipedia.org/wiki/Windrow_composting?oldid=588333818 *Contributors:* Shoehorn~enwiki, Tsavage, Cyrius, Pengo, Alan Liefting, Rparle, PDH, Remuel, ·~enwiki, Vortexrealm, Maurreen, Jwanders, Kollision, Splash, SmackBot, Radagast83, Hebrides, Alphachimpbot, Jc2it, CommonsDelinker, Plymouths, Red58bill, Amornoguerra, Psychless, Jefflayman, Debresser, Lightbot, Fsims, Thomas swoon, Duboshi, Citation bot 1, Lotje, Fox Wilson, Difu Wu, Look2See1, BG19bot, Raperroni, The Illusive Man, BurritoBazooka and Anonymous: 17

39.6.2 Images

- **File:A_small_cup_of_coffee.JPG** *Source:* https://upload.wikimedia.org/wikipedia/commons/4/45/A_small_cup_of_coffee.JPG *License:* CC BY-SA 2.0 *Contributors:* Own work *Original artist:* Julius Schorzman

- **File:Aeration_floor.jpg** *Source:* https://upload.wikimedia.org/wikipedia/commons/2/2f/Aeration_floor.jpg *License:* CC-BY-SA-3.0 *Contributors:* Transferred from en.wikipedia to Commons. *Original artist:* The original uploader was LLucas at English Wikipedia

- **File:Ambox_globe_content.svg** *Source:* https://upload.wikimedia.org/wikipedia/commons/b/bd/Ambox_globe_content.svg *License:* Public domain *Contributors:* Own work, using File:Information icon3.svg and File:Earth clip art.svg *Original artist:* penubag

- **File:Ambox_important.svg** *Source:* https://upload.wikimedia.org/wikipedia/commons/b/b4/Ambox_important.svg *License:* Public domain *Contributors:* Own work, based off of Image:Ambox scales.svg *Original artist:* Dsmurat (talk · contribs)

- **File:Ambox_style.png** *Source:* https://upload.wikimedia.org/wikipedia/en/d/d6/Ambox_style.png *License:* ? *Contributors:*
  Derived from Image:Broom icon.svg, which was copied from http://www.kde-look.org/content/show.php?content=29699 *Original artist:*
  Created by gg3po (Tony Tony) as an SVG, copied to Wikimedia Commons by Booyabazooka, converted to PNG with tweaked transparency for older web browsers by David Levy.

- **File:Anisakids.jpg** *Source:* https://upload.wikimedia.org/wikipedia/commons/d/d2/Anisakids.jpg *License:* Public domain *Contributors:* Originally from en.wikipedia; description page is/was en:Image:Anisakids.jpg. *Original artist:* Anilocra

- **File:Anthelmintic_effect_of_papain_on_Heligmosomoides_bakeri.ogv** *Source:* https://upload.wikimedia.org/wikipedia/commons/f/f1/Anthelmintic_effect_of_papain_on_Heligmosomoides_bakeri.ogv *License:* CC BY 2.0 *Contributors:* Behnke, J. M.; Buttle, D. J.; Stepek, G.; Lowe, A.; Duce, I. R. (2008). "Developing novel anthelmintics from plant cysteine proteinases". Parasites & Vectors 1: 29. doi: 10.1186/1756-3305-1-29. *Original artist:* Behnke, J. M.; Buttle, D. J.; Stepek, G.; Lowe, A.; Duce, I. R. (2008). "Developing novel anthelmintics from plant cysteine proteinases". Parasites & Vectors 1: 29. doi:10.1186/1756-3305-1-29.

- **File:Ants_cleaning_dead_snake.jpg** *Source:* https://upload.wikimedia.org/wikipedia/commons/3/30/Ants_cleaning_dead_snake.jpg *License:* CC-BY-SA-3.0 *Contributors:* ? *Original artist:* ?

- **File:Bedrijfsafval.jpg** *Source:* https://upload.wikimedia.org/wikipedia/commons/3/3a/Bedrijfsafval.jpg *License:* Public domain *Contributors:* ? *Original artist:* ?

- **File:Black_soldier_flies_mating.jpg** *Source:* https://upload.wikimedia.org/wikipedia/commons/f/fe/Black_soldier_flies_mating.jpg *License:* GFDL *Contributors:* Own work *Original artist:* **Muhammad Mahdi Karim** (www.micro2macro.net) Facebook Youtube

- **File:Black_soldier_fly_depositing_eggs_in_cardboard.jpg** *Source:* https://upload.wikimedia.org/wikipedia/commons/7/72/Black_soldier_fly_depositing_eggs_in_cardboard.jpg *License:* CC BY-SA 3.0 *Contributors:* Own work *Original artist:* blacksoldierflyblog.com

- **File:Black_soldier_fly_inflating_wings.jpg** *Source:* https://upload.wikimedia.org/wikipedia/en/6/60/Black_soldier_fly_inflating_wings.jpg *License:* CC-BY-SA-3.0 *Contributors:*
  Photographed a single adult black soldier fly over a 15 minute time frame and combined them.
  Previously published: My site: http://blacksoldierflyblog.com/ *Original artist:*
  Gee W.

39.6.3   Content license

www.ingramcontent.com/pod-product-compliance
Lightning Source LLC
Chambersburg PA
CBHW080808180526

45168CB00006B/2364